T0135793

High temperature polymer electrolyte membrane fuel cells –

Modeling, simulation, and segmented measurements

Von der Fakultät für Ingenieurwissenschaften, Abteilung Maschinenbau der

Universität Duisburg-Essen

zur Erlangung des akademischen Grades

DOKTOR-INGENIEUR

genehmigte Dissertation

von

Christian Siegel

aus

Mondercange, Luxemburg

Referent: Prof. Dr. rer. nat. Angelika Heinzel

Korreferent: Prof. Dr.-Ing. Stefan Pischinger

Tag der mündlichen Prüfung: 23.01.2015

Bibliografische Information der Deutschen Nationalbibliothek

Die Deutsche Nationalbibliothek verzeichnet diese Publikation in der
Deutschen Nationalbibliografie; detaillierte bibliografische Daten sind
im Internet über http://dnb.d-nb.de abrufbar.

ISBN 978-3-8325-3917-7

Logos Verlag Berlin GmbH
Comeniushof, Gubener Str. 47,
10243 Berlin
Tel.: +49 (0)30 42 85 10 90
Fax: +49 (0)30 42 85 10 92
INTERNET: http://www.logos-verlag.de

Abstract

A complete three-dimensional (3D) model of a high temperature polymer electrolyte membrane (HTPEM) fuel cell, employing a high temperature stable polybenzimidazole membrane electrode assembly (MEA of approximately 50 cm^2) doped with phosphoric acid (PBI/H$_3$PO$_4$), was developed and implemented using a commercially available finite element software. The model provided a one-to-one representation of the HTPEM fuel cell that was used in all of the experimental tests. Three types of flow-fields (6 channel parallel serpentine flow-field, parallel straight flow-field, and mixed-type flow-field) were modeled and simulated. The model included the conservation equations of mass, momentum, species, charge, and energy, which were solved in their respective subdomains. The agglomerate model for the electrochemical reactions separately accounts for the transport resistance of the species inside the spherical agglomerate, the amorphous phase phosphoric acid film, and the phosphoric acid water mixture film. A new, two-phase temperature approach was used to separately address the fluid-(gas)-, and the solid-phase temperature distributions. All boundary equations were based on the experimental set-up. After meshing the model geometry and solving the model with a dedicated solution procedure, selected simulation results at reference operating conditions were compared to the performance curves and to segmented solid-phase temperature and current density measurements. The computational fluid dynamic (CFD) results supported the observed trends in terms of the fluid-flow distribution. Overall, similar behaviours were observed for the simulation results and the experimental results. For the segmented measurements, an inhouse developed HTPEM fuel cell (3 flow-fields with 36 segments each) was designed and manufactured. The HTPEM fuel cell was heated-up from the bottom side only (anode side facing down and cathode side facing up in the experimental set-up). The bipolar-plates and the membrane electrode assembly were sandwiched between two polyetheretherketone (PEEK) plates to minimize the influence of the heating elements on the solid-phase temperature distribution. The HTPEM fuel cell was successfully operated in a teststand built inhouse, and the solid-phase temperature and the current density distribution were recorded, evaluated, and discussed. An influence of the fluid-(gas)-phase temperature on the solid-phase temperature was reported for different flow rates at no-load operating conditions. At load operating conditions, the highest solid-phase temperature was observed in the area of the highest current density, which was typically close to the cathode inlet. The current density distribution was primarily dependent on the oxygen availability and the fluid-flow distribution. The current density distribution was fairly stable against changes in the anode stoichiometric flow rate whereas changes in the cathode stoichiometric flow rate exerted a significant impact. The trend changed when carbon monoxide (CO) enriched hydrogen was used at the anode side. In this case, the highest current density tended to shift towards the anode inlet and significantly overlapped the shape given by the oxygen availability. Therefore, when using CO enriched hydrogen, the counter-flow configuration produced a flatter current density distribution than the co-flow configuration. These measurements were performed for CO contents up to several percent, and the effect was less pronounced for a low CO percentage. Additionally, the solid-phase temperature distribution changed according to the CO content. During all tests, a slightly higher solid-phase temperature was measured when using CO enriched hydrogen compared to the same operating conditions with pure hydrogen. For the first time and for selected operating conditions, segmented electrochemical impedance spectroscopy (EIS) measurements were performed to qualitatively support the observed trends. These measurements were used to identify and determine the causes of the inhomogeneous current density distributions. An equivalent circuit model was developed, the obtained spectra were analyzed, and the model parameters discussed. For a HTPEM fuel cell operating with hydrogen and air, a reduced low frequency resistance was measured at segments close to the cathode inlet in combination with an increased low frequency resistance at segments close to the cathode outlet. These results were in accordance with the fluid-flow distribution and the oxygen availability. When the HTPEM fuel cell was operated with CO enriched hydrogen, an increased high frequency resistance was confirmed at segments close to the anode outlet along with a reduced high frequency resistance at segments close to the anode inlet. It was the first time that segmented measurements were performed to this extent for a HTPEM fuel cell. This work helps to provide a better understanding of the internal behaviour of a running HTPEM fuel cell and presents valuable data for modeling and simulation. For large fuel cells and complete fuel cell stacks in particular, well designed anode and cathode inlet and outlet sections are expected to aid in achieving flatter quantities distributions and in preventing hot spots over the membrane electrode assembly area. Finally, the reported solid-phase and current density distributions may be useful in the development of proper start-up, shut-down, and tempering concepts for HTPEM fuel cells and stacks.

Zusammenfassung

In dieser Arbeit wurde ein dreidimensionales (3D), strömungsmechanisches Modell einer Hochtemperatur-Polymer-Elektrolyt-Membran-(HTPEM)-Brennstoffzelle entwickelt und in eine kommerziell erhältliche finite-Elemente-Software implementiert. Das Modell beinhaltet eine hochtemperaturstabile Polybenzimidazol-(PBI)-Membran-Elektroden-Einheit (MEA), welche mit Phosphorsäure dotiert ist (PBI/H_3PO_4). Die verwendete Modellgeometrie ist eine exakte Kopie der HTPEM-Brennstoffzelle, welche in allen experimentellen Tests eingesetzt wurde. Insgesamt wurden drei verschiedene Kanalstrukturen theoretisch und experimentell unter verschiedenen Betriebsbedingungen untersucht (6-Kanal-Mäanderstruktur, 26-Kanal-Parallelstruktur, gemischte Kanalstruktur mit teilweise Serpentinen-, teilweise Parallelstruktur). Das entwickelte Modell wurde mittels Erhaltungsgleichungen beschrieben. Ein Zweiphasen-Temperaturmodell wurde angesetzt, um die Fluidphasen- und Festkörpertemperatur sowie deren Wechselwirkung zu beschreiben. Die Randbedingungen des Modells wurden exakt nach den Vorgaben der experimentellen Tests definiert. Anschließend wurde die komplette Modellgeometrie diskretisiert und das Modell numerisch gelöst. Die Simulationsergebnisse wurden für die Referenzbedingungen angegeben und mit der Strom- und Spannungskennlinie sowie mit segmentierten Festkörpertemperatur- sowie Stromdichtemessungen verglichen. Strömungsmechanische Simulationen unterstützten diese Ergebnisse bezüglich der Strömungsverteilung und des Druckverlustes für alle drei Kanalstrukturen. Insgesamt wurde eine gute Übereinstimmung zwischen den theoretischen und den experimentellen Ergebnissen nachgewiesen. Für die segmentierten Messungen wurde eine spezielle HTPEM-Brennstoffzelle entwickelt (jeweils 36 Segmente pro Kanalstruktur). Die gesamte Messzelle wurde von der Anodenseite aus mittels zwei Heizelementen auf Temperatur gebracht und gehalten. Die Membran-Elektroden-Einheit sowie die Bipolarplatten (nicht segmentierte Anodenseite und segmentierte Kathodenseite) wurden zwischen Polyetheretherketon-(PEEK)-Platten mit niedriger thermischer Leitfähigkeit eingebaut, um die segmentierte Festkörpertemperaturmessung nicht durch die Heizelemente zu verfälschen. Die segmentierte HTPEM-Brennstoffzelle wurde erfolgreich unter Laborbedingungen getestet und segmentierte Stromdichte- und Festkörpertemperaturmessungen aufgenommen. Bei den segmentierten Festkörpertemperaturmessungen wurde ein Einfluss der Fluidphasentemperatur auf die Festkörpertemperatur für unterschiedliche Betriebsbedingungen nachgewiesen. Dieser Einfluss war im stromlosen Betriebszustand besonders ausgeprägt. Im Belastungsfall wurde die höchste Festkörpertemperatur im Bereich der höchsten Stromdichte gemessen, welche in den meisten Fällen im Bereich des Kathodeneingangs vorzufinden war. Die Stromdichteverteilung ist im Wesentlichen von der Sauerstoffverteilung und den Strömungsverhältnissen abhängig. Des Weiteren wurde nachgewiesen, dass die Stromdichteverteilung sehr stabil gegenüber Änderungen der Anodenstöchiometrie ist und dass Änderungen der Kathodenstöchiometrie einen wesentlichen Einfluss auf die Verteilung haben. Das Verhalten der Stromdichteverteilung ändert sich, wenn kohlenmonoxidhaltiger Wasserstoff verwendet wird. In diesem Falle steigt die Stromdichte im Bereich des Anodeneingangs an und fällt im Bereich des Anodenausgang ab. Der Einfluss der Sauerstoffverteilung auf die Stromdichte ist nach wie vor vorhanden, wird aber durch diese Stromdichteverschiebung überdeckt. Der Einfluss steigt mit steigendem Kohlenmonoxidanteil im Wasserstoff, so dass die höchste Stromdichte unter Umständen im Bereich des Anodeneingangs zu finden ist. In diesem Sinne ändert sich auch die Verteilung der Festkörpertemperatur. Bedingt durch diesen Sachverhalt ist es vorteilhaft, die HTPEM-Brennstoffzelle in Gasgegenstromverschaltung zu betreiben. Es werden gleichmäßigere Stromdichteverteilungen erreicht als bei einer Gasgleichstromverschaltung. Des Weiteren wurden im Betrieb mit kohlenmonoxidhaltigem Wasserstoff leicht erhöhte Festkörpertemperaturen gemessen. Für einige ausgewählte Betriebsbedingungen wurden segmentierte elektrochemische Impedanzspektroskopiemessungen (EIS) als unterstützende Maßnahme eingesetzt. Folglich wurde ein Ersatzschaltbild für die HTPEM-Brennstoffzelle entwickelt und die Spektren analysiert. Die Modellparameter unterstützten die segmentierten Stromdichtemessungen hinsichtlich der erhöhten Transportwiderstände im Bereich des Kathodenausgangs für einen Betrieb mit Luft und Wasserstoff. Der Einfluss der Kanalstruktur auf die lokalen Spektren wurde ebenfalls nachgewiesen. Das Gleiche gilt hinsichtlich erhöhter Transportwiderstände im Bereich des Anodenausgangs für einen Betrieb mit Luft und kohlenmonoxidhaltigem Anodengas. Die vorliegende Arbeit dient zum besseren Verständnis der internen Vorgänge einer HTPEM-Brennstoffzelle im Betrieb. Sie soll einen Beitrag leisten, die Ergebnisse von HTPEM-Brennstoffzellenmodellen besser bewerten zu können. Dies ist insbesondere für HTPEM-Brennstoffzellenstacks interessant, da hier wesentlich höhere Volumen- und Temperiermedienströme verwendet werden müssen. Schlussendlich kann man mit einem verbesserten Design der Bipolarplatten und der Kanalstrukturen sowie der geschickten Platzierung von anoden- und kathodenseitigen Ein- und Auslässen gleichmäßigere Verteilungen erreichen und dadurch mögliche Hot-Spots vermeiden, was sich positiv auf die Langlebigkeit der HTPEM-Brennstoffzelle auswirken kann.

Acknowledgments

I completed my diploma thesis on low temperature polymer electrolyte membrane fuel cell modeling in 2004 at the Karlsruhe Institute of Technology (KIT – former University of Karlsruhe TH), and then I decided to continue my research in fuel cells in 2007. Over the last few years I have had the opportunity to work at the Centre for Fuel Cell Technology (Zentrum für BrennstoffzellenTechnik – ZBT) in Duisburg, Germany. The present dissertation is the outcome of my research in high temperature polymer electrolyte membrane fuel cells.

First, I would like to thank my Prof. Dr. rer. nat., Angelika Heinzel, for giving me the opportunity to pursue my research, for supervising me, and for her encouragement and endurance over the last years. I am thankful for our many discussions, her helpful comments on the text, her correction of this manuscript, and her support in various publications. I am very grateful to Prof. Dr.-Ing. Stefan Pischinger from the Institute for Combustion Engines VKA, RWTH Aachen University, Germany for being my second supervisor. I would also like to thank my dissertation committee members for participating in this process.

Several individuals helped me during my time at the Centre for Fuel Cell Technology. I am grateful to the head of the Fuel Cells & Systems Department, Dr. Peter Beckhaus, for his professional support and helpful suggestions and to Dr. Jens Burfeind for excellent scientific feedback and fruitful discussions. I especially thank Dr. George Bandlamudi for teaching me how to work in the laboratory and how to handle fuel cells and for active cooperation during all experimental tests. I appreciate all of the staff members from the laboratory and the workshop for their individual support and feedback during my experimental work. I also thank all other individuals at the Centre for Fuel Cell Technology for making my time a rewarding experience through either technical discussions or private communication.

I appreciate my family and Lynn Schwachtgen for supporting me over the last years.

I am grateful to the Fonds National de la Recherche Luxembourg for initiating a project in which I had the opportunity to work as a PhD student. This work was supported by 'LE GOUVERNEMENT DU GRAND- DUCHÉ DE LUXEMBOURG, MCESR Recherche et Innovation', Grant No.: AFR07/007.

Table of content

Nomenclature

a	Activity	
	Surface area	m^2
c	Concentration	$mol\ m^{-3}$
	General constant	
C	Capacity, specific	$F, F\ m^{-2}$
C_p	Heat capacity	$J\ kg^{-1}\ K^{-1}$
d	Average hopping distance between sites	m
D	Diffusion coefficient	$m^2\ s^{-1}$
E	General cell potential	V
E_{cell}	Cell potential at an arbitrary operating point	V
E_T	Reversible cell potential at an arbitrary temperature	V
E^0	Theoretical maximum cell potential	V
ΔE^a	Activation energy	$J\ mol^{-1}$
f	Frequency	Hz
	Ratio factor	
F	Faraday constant	$A\ s\ mol^{-1}$
G	Gibbs free energy	$J, J\ mol^{-1}$
h	Heat transfer coefficient (area, volume)	$W\ m^{-2}\ K^{-1}, W\ m^{-3}\ K^{-1}$
H	Enthalpy	$J, J\ mol^{-1}$
	Solubility	$mol\ m^{-3}\ atm^{-1}\ (Pa^{-1})\ (bar^{-1})$
I	Identity	
j	Current density	$A\ m^{-2}$
	Imaginary unit, sqrt(-1)	
J	General radiosity	$W\ m^{-2}$
k	Thermal conductivity	$W\ m^{-1}\ K^{-1}$
k_c	Reaction rate constant	s^{-1}
k_N	Relative reach of diffusion compared to a finite length	s^{-1}
k_p	Permeability	m^2
l	General thickness	m
L	Length	m
	Inductance	H
L_{entr}	Entry length	m
m	Loading	$kg\ m^{-2}$

M	Molar mass	kg mol^{-1}
n	General number	
p	Pressure	Pa
p_{entr}	Pressure at gas channel entry – weak contribution	Pa
P	Power	W
P_i	Element type	
q_0	Heat flux	W m^{-2}
Q	Charge	C
r	Radius	m
R	Gas constant	J mol^{-1} K^{-1}
	Resistance	Ω
S	Entropy	J K^{-1}, J mol^{-1} K^{-1}
	General source or sink term	kg m^{-3} s^{-1}, A m^{-3}, W m^{-3}
t	Time	s
T	General temperature	°C, K
u	Velocity vector	m s^{-1}
U	Internal energy	J, J mol^{-1}
	Voltage	V
U_{cell}	Cell voltage at an arbitrary operating point	V
V	Volume	m^3
W	Warburg parameter	Ω s$^{-1/2}$
	Work term	J, J mol^{-1}
x	Dimension along the x-axis	m
	Mole fraction	
X	Membrane doping level	
y	Dimension along the y-axis	m
z	Charge number	
	Dimension along the z-axis	m
Z	General impedance, specific	Ω, Ω m^2

Greek letters

α	Transfer coefficient	
α_h	Reciprocal of the number of all possible hopping directions	
γ	Exponent factor	
Γ	Characteristic length scale	m

δ	Layer thickness	m
ε	Volume fraction	
η	Dynamic viscosity	Pa s
	Efficiency	
	Overpotential	V
λ	Stoichiometry	
μ	Chemical potential	J, J mol^{-1}
v	Diffusion volume	m^3 mol^{-1}
	Stoichiometric coefficient	
v_0	Hopping frequency	
ξ	General variable	
ρ	Density	kg m^{-3}
σ	Conductivity	S m^{-1}
	Pre-exponential factor	S K m^{-1}
φ	Phase shift of response signal	
ϕ	Phase potential	V
ϕ_L	Thiele modulus	
ψ	Lagrange multiplier	
ω	Mass fraction	
	Radial frequency	Hz, rad s^{-1}

Superscripts

0	Reference, standard, void
agg	Agglomerate
eff	Effective
f	Formation
g	Gas
s	Solid-phase
T	Transposed

Subscripts

0	Reference, standard, void
a	Anode side
act	Activation
$boundary$	Value at respective boundary
c	Capacitance

	Cathode side
conc	Concentration
Cu	Copper
dl	Double layer
eff	Effective
el	Electrical
f	Fluid-(gas)-phase
fuel	Efficiency factor due to gas feed
HF	High frequency
i	Index
ionic	Ionic conduction
IM	Imaginary part
j	Index
k	Index
L	Limiting
	Inductance
LF	Low frequency
N	Nernst
OCV	Open circuit voltage (potential)
Ohmic (Ω)	Ohmic
prod	Product
PE	Porous electrode
Pt/C	Platinum-to-carbon ratio
react	Reactant
ref	Reference
RE	Real part
s	Solid-phase
th	Theoretical
tot	Total
T_s/T_f	Solid to fluid heat transfer
voltage	Efficiency factor due to cell operation
W	Warburg

Chemical abbreviations

C	Carbon
CO	Carbon monoxide

CO_2	Carbon dioxide
CO_3^{2-}	Carbonate
e^-	Electron
H^+	Proton
H_2	Hydrogen
H_2O	Water
H_3PO_4	Phosphoric acid (PA)
$H_4P_2O_7$	Pyrophosphoric acid (PPA)
N_2	Nitrogen
O_2	Oxygen
O^{2-}	Oxide ion
OH^-	Hydroxide
PBI	Polybenzimidazole
$PBI/(X-2)H_3PO_4$	PBI/amorphous phase H_3PO_4
Pt	Platinum

General abbreviations

1D, 2D, 3D	One-, two-, three-dimensional
AFC	Alkaline fuel cell
BPP	Bipolar-plate
CAD	Computer aided design
CFD	Computational fluid dynamics
CHP / μCHP	(Micro) combined heat and power
CU	Gold-plated copper current collector
DC	Direct current
DMAc	Dimethylacetamide
DOF	Degrees of freedom
EIS	Electrochemical impedance spectroscopy
FEM	Finite element method
FKM	Fluoroelastomer
GDL	Gas diffusion layer
GMRES	Generalized minimal residual method
HOR	Hydrogen reduction reaction
HTPEM	High temperature polymer electrolyte membrane fuel cell
I-V-curve	Current-voltage-curve
KOH	Potassium hydroxide

LTPEM	Low temperature polymer electrolyte membrane fuel cell
MCFC	Molten carbonate fuel cell
MEA	Membrane electrode assembly
MEM	Membrane
OCV	Open curcuit potential
ORR	Oxygen reduction reaction
PAFC	Phosphoric acid fuel cell
PDE	Partial differential equation
PEEK	Polyetheretherketone
PEM	Polymer electrolyte membrane
PEMFC	Polymer electrolyte membrane fuel cell
PFSA	Perfluorosulfonic
PID	Proportional integral derivative controller
PIV	Particle image velocimetry
PTFE	Polytetrafluoroethylene
RAM	Random access memory
RC	Resistor capacitor element
RL	Reaction layer
RTD	Resistance temperature detector
SiC	Silicon carbide
SOFC	Solid oxide fuel cell
SOR(U)	Successive over-relaxation (preconditioner)
TFA	Trifluoroacetic acid

1. Introduction

The dependency on fossil fuels as a primary energy source must be reduced, and suitable renewable substitutes must be found. Among others, hydrogen technology is expected to play an important role as a clean energy carrier over the next few decades to meet the strict energy and carbon saving criteria. The energy sector is a major economic factor and advanced hydrogen technology such as water electrolysis, solar hydrogen generation, and energy storage technology, will create new business opportunities [1]. Highly efficient fuel cell based systems for portable, transport, and stationary applications appear to be an attractive alternative to achieve less pollution, a secure energy supply, a cleaner environment, and sustainability. Fuel cells are electrochemical power sources that directly convert the energy stored in a fuel into electricity and heat. These cells operate without a combustion chamber and thus, significantly differ from traditional heat engines by the fact that their efficiency is not limited by the Carnot cycle and that high overall efficiency can be achieved at a much lower operating temperature. Among the numerous types of fuel cells, the polymer electrolyte membrane (PEM) fuel cell is a promising candidate for future energy systems. Low temperature PEM (LTPEM) fuel cells operate at approximately 80°C and are a major field of interest in the green energy community. Single cells, fuel cell stacks, and complete systems have been continuously improved over the years. Currently, prototype mobile applications exist, fuel-cell powered demonstration vehicles have been developed, and field tests of (micro) combined heat and power units ((μ)CHP) are being pursued in countries around the globe to prepare for commercialization, particularly in Germany, Japan and South-Korea (see, e.g., [2-4]). These systems are expected to be ready for installation in the next few years to deliver power and heat for single family households [5-7]. As stated in [8], 2011 was by far the most successful year to date in the history of fuel cells, with the annual megawatts shipped exceeding 100 MW for the first time, as commercialization of the industry took hold. To overcome some of the drawbacks directly related to the low operating temperature, the possibility of operating PEM fuel cells at a higher operating temperature was proposed over 15 years ago [9-11]. Polybenzimidazole (PBI) doped with phosphoric acid (H_3PO_4) has been proposed as a proton conducting membrane that can operate at 160°C. Extensive research and development efforts have been conducted to continuously improve the performance, durability, and long-term stability of the PBI/H_3PO_4 membrane electrode assembly (MEA). In high temperature PEM (HTPEM) fuel cells, water flooding and water management is less critical, the cooling process is simpler, carbon monoxide (CO) poisoning of the platinum (Pt) catalyst is less prominent, and the electrode kinetics are generally faster compared to LTPEM fuel cells. Today, PBI/H_3PO_4 membrane electrode assemblies and HTPEM fuel cell based systems are commercially available (see, e.g., [12,13]). Nevertheless, the improvement of all components and general cost reductions are necessary to make this technology more attractive and cost competitive with current energy conversion devices. As HTPEM fuel cell technology is relatively new, several questions related to the layout of key cell components still persist. One reason for these questions is that there is a lack of information available on the internal behaviour of a working HTPEM fuel cell. To continuously improve the layout of the key components and to adapt operating schemes as controlled start-up and shut-down procedures, quantities distributions for various operating conditions are indispensable. Modeling and simulation of HTPEM fuel cells represent a new and important tool that can provide additional understanding. In this work, a complete 3D model of a HTPEM fuel cell was developed, implemented, and solved using finite element software. All conservation equations were solved in the respective subdomains. A two-phase temperature approach was used to separately account for the fluid-(gas)-, and the solid-phase temperature distributions. The simulation results are discussed for three types of flow-fields (6 channel parallel serpentine flow-field, parallel straight flow-field, and mixed-type flow-field) and are compared to the overall performance curves and to segmented measurements. An inhouse developed HTPEM fuel cell with an exchangeable segmented cathode side flow-field was employed to measure the solid-phase temperature and current density distribution. For a few selected operating conditions, segmented EIS measurements were performed to qualitatively support the observed trends. These measurements helped to identify the causes of inhomogeneous current density distributions, which can create hot spots over the membrane electrode assembly area, possibly causing local degradation. Some of the obtained spectra were analyzed using an equivalent circuit model, and the model parameters discussed.

2. Fuel cells

The first generation of electrical energy by electrochemical conversion of hydrogen and oxygen was achieved over 160 years ago. William Robert Grove was the first to report the successful operation of a hydrogen-oxygen gas battery [14,15]. However, it appears that a Swiss scientist independently discovered the same effect at approximately the same time [16]. In the following years, various types of gas batteries and cells were more or less successfully presented, including the solid oxide fuel cell (SOFC) and the molten carbonate fuel cell (MCFC), both operating at a considerably higher temperature. Researchers subsequently developed new types of fuel cells, including the alkaline fuel cell (AFC) and the phosphoric acid fuel cell (PAFC). The first practical fuel cell applications arose in the early 1960s as an integral part of several space programs.

Table 1

Major five types of fuel cells, typical operating temperature, and involved electrochemical reactions.

Type	T_s	Anode side reaction	Cathode side reaction
	°C		
AFC	60-200	$H_2 + 2OH^- \rightarrow 2H_2O + 2e^-$	$\frac{1}{2}O_2 + H_2O + 2e^- \rightarrow 2OH^-$
PEMFC	50-80	$H_2 \rightarrow 2H^+ + 2e^-$	$\frac{1}{2}O_2 + 2H^+ + 2e^- \rightarrow H_2O$
PAFC	150-200	$H_2 \rightarrow 2H^+ + 2e^-$	$\frac{1}{2}O_2 + 2H^+ + 2e^- \rightarrow H_2O$
MCFC	640-660	$H_2 + CO_3^{2-} \rightarrow H_2O + CO_2 + 2e^-$	$\frac{1}{2}O_2 + CO_2 + 2e^- \rightarrow CO_3^{2-}$
SOFC	800-1,000	$H_2 + O^{2-} \rightarrow H_2O + 2e^-$	$\frac{1}{2}O_2 + 2e^- \rightarrow O^{2-}$

Some 40 years ago, the Nafion® based polymer was introduced as a membrane material for the PEM fuel cell. Today, there are five major types of fuel cells with a nearly identical operating principle. Their classification is based on the operating temperature (solid-phase temperature), which dictates the physical, chemical, and mechanical properties of the used materials [1]. Additionally, fuel cells can be categorized based on the electrolyte that separates the anode side from the cathode side. The electrolyte dictates the nature of the ions that are transferred, the direction of the transport, and the side on which water is produced [17]. Table 1 lists typical operating temperatures and the electrochemical reactions involved for the five major types of fuel cells. AFCs can operate over a large temperature window of 60-200°C depending on the concentration of the electrolyte (KOH). In AFCs, the nature of the transferred ion is OH^-. This type of fuel cell offers the potential for using non-precious metal catalysts whereas a drawback is that it is intolerant to carbon dioxide. Pure hydrogen and oxygen must be used during operation. SOFCs and MCFCs operate at a considerably higher temperature. The nature of the transferred ion are O^{2-} and CO_3^{2-} respectively. In these types of fuel cells, the fuel flexibility is greatly enhanced due to the high operating temperature and the ability to use non-precious metal catalysts. The PAFC uses phosphoric acid as an electrolyte, while operating at a temperature of 150-200°C. The electrolyte is contained in a silicon carbide (SiC) matrix between two porous coated electrodes with a platinum catalyst. In PAFCs, protons are transferred through the electrolyte. PEMFCs have drawn much attention due to their simple operation at low temperatures. Compared to liquid electrolyte based systems, polymer electrolyte systems are easier to handle, resulting in a relatively simple construction. The polymer membrane is coated on both sides with a platinum catalyst and porous carbon electrode support. Because the membrane must be hydrated with water to maintain adequate proton conductivity, its operating temperature is limited to 80°C. LTPEM fuel cells can only tolerate a low amount of CO in the anode gas, whereas fuel cells operating at a higher temperature are less sensitive to fuel contaminants. Thus, it follows that the HTPEM fuel cells can combine several advantages of the PAFC and LTPEM fuel cell technology, as will be discussed below. A more detailed overview of the different types of fuel cells can be found in, e.g., [17-20].

2.1 Working principle

The working principle of a PEM fuel cell is described in many textbooks [17-20]. At the anode side of the cell, fuel is fed into the gas channels of the bipolar-plate (BPP). The gas diffuses through a gas diffusion layer (GDL) made out of carbon cloth or carbon paper towards the reaction layer (RL), that is, the catalyst layer or electrode. At the anode, the electrochemical half-cell reaction takes place at the triple phase boundary. A common catalyst for both reactions is platinum, whereas the catalyst support material is typically carbon powder (Vulcan XC-72). The hydrogen splits into protons and electrons. The anode side half-cell reaction is given by Eq.(1) and is known as the hydrogen oxidation reaction (HOR).

$$2 \cdot H_2 \rightarrow 4 \cdot H^+ + 4 \cdot e^- \tag{1}$$

The polymer electrolyte membrane is sandwiched between the anode and cathode side. It is an insulator for the electrons and is ideally impermeable to gases. The membrane exhibits a relatively high proton conductivity that is dependent on the membrane structure itself and the water content. Protons are transferred through the membrane in the form of hydronium ions. Due to the prevailing potential difference, the electrons travel via an external load, where power can be drawn, to the cathode side reaction layer. At the cathode side of the cell, air (oxygen) is fed into the cell and reacts with the protons coming from the membrane and the electrons coming from the external load. Eq.(2) represents the cathode side electrochemical half-cell reaction, known as the oxygen reduction reaction (ORR).

$$O_2 + 4 \cdot H^+ + 4 \cdot e^- \rightarrow 2 \cdot H_2O \qquad (2)$$

Both electrochemical half-cell reactions occur simultaneously at both reaction layers. According to Eq.(3), water is produced at the cathode side and, thermal power is released due to the exothermic reactions.

$$2 \cdot H_2 + O_2 \rightarrow 2 \cdot H_2O \qquad (3)$$

The fuel cell electrochemically converts the energy stored in a fuel into electrical power by the oxidation of hydrogen at the anode side and the reduction of oxygen at the cathode side. Fig.1 depicts the principle of operation for a PEM fuel cell, along with various components.

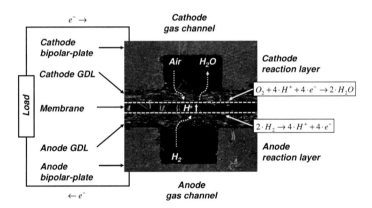

Fig.1. Principle of operation of a PEM fuel cell along with the different involved components and electrochemical half-cell reactions (y-z-plane).

2.2 HTPEM fuel cells

Polymer electrolyte membranes, such as Nafion®, are characterized by chemical stability, mechanical strength, and low gas permeability. The operating temperature is limited to 80°C to maintain a high water content and, thus a high proton conductivity and a good performance [17-20]. To overcome the challenges that are directly associated with the low operating temperature, the possibility of using a higher operating temperature was proposed several years ago. The properties of the currently used perfluorosulfonic (PFSA) membranes had to be modified and new membrane materials had to be developed. Several classes of membrane materials that can operate at a higher temperature have been investigated in the past (e.g., copolymer, non-fluorinated, and alternative liquid electrolytes) [21-26]. Moreover, several groups have studied phosphoric acid doped membranes based on Nafion®, PBI, and their composites [27-32].

2.3 PBI/H$_3$PO$_4$ MEA

Savinell et al. [33] published the first studies on the conductivity of Nafion® equilibrated with concentrated phosphoric acid. This membrane was operated at a higher temperature and exhibited a proton conductivity of 2 S m^{-1} under a dry atmosphere and 5 S m^{-1} at 20% relative humidity. The high performance polymer PBI received considerable attention as an alternative membrane material due to its reported glass transition temperature of 430°C [34]. The possibility of using a PBI/H$_3$PO$_4$ membrane as a proton conductor in fuel cells was introduced by Wainright et al. [9], Samms et al. [10], and Wang et al. [11]. These authors tested PBI films doped with phosphoric acid as potential polymer electrolytes for fuel cell applications. Over the last decades, various membrane casting methods have been investigated for, including TFA (mixture of phosphoric acid and trifluoroacetic acid) cast membranes and DMAc (with an organic solvent of N,N-dimethylacetamide) cast membranes [35,36]. Xiao et al. [37] presented a sol-gel process to produce PBI/H$_3$PO$_4$ membranes. Polyphosphoric acid (PPA) was used as an efficient condensation reagent and as a solvent for the PBI synthesis. After casting, the hydrolysis of polyphosphoric acid to phosphoric acid by moisture from the surrounding environment induced a sol-to-gel transition, resulting in highly doped PBI/H$_3$PO$_4$ membranes. The achieved doping levels were as high as 40 mol of phosphoric acid per PBI repeat unit, resulting in proton conductivity values exceeding 20 S m^{-1}. Today, highly doped PBI/H$_3$PO$_4$ sol-gel membrane electrode assemblies are commercialized by BASF fuel cells under the brandname Celtec P®-Series [38]. Other types of commercially available HTPEM fuel cell membranes from, e.g., FuMA-Tech [39], or Advent [40] are discussed in [41].

2.4 Overall features of a PBI/H$_3$PO$_4$ MEA

In addition to its heat resistance, PBI is chemically and mechanically stable, and has a low gas permeability. In its pure form, PBI is an excellent electrical and ionic insulator and can take up large quantities of phosphoric acid to achieve reasonable levels of proton conductivity. The doping level of these membranes is defined as the number of phosphoric acid molecules per PBI repeat unit. Once the phosphoric acid is adsorbed, it can be distinguished as chemically bonded acid or non-bonded or free acid. The chemically bonded acid forms bonds with the N-H groups of the PBI. A maximum of two equivalents of acid are bonded to each PBI repeat unit. For doping levels higher than two, some of the acid will remain non-bonded or free [1]. It has been suggested that amorphous phase phosphoric acid is primarily responsible for the proton transport. The proton transfer mainly occurs by the Grotthuss mechanism, in which proton are hopping through the hydrogen bond network of acid and water molecules through the formation of covalent bonds [1,42]. There is no net translation of the electrolyte species. The proton conductivity of the doped material is nearly as high as that of perfluorosulfonic membranes but is far less dependent on the relative humidity. Thus, dry gases can be used, leading to a significant reduction of complexity and costs, as humidification of the gases is not needed. Regarding PAFCs, the electrolyte system in a PBI/H_3PO_4 membrane electrode assembly is essentially solid and, therefore, is easier to handle, is more tolerant to pressure variation, and raises fewer difficulties concerning the electrolyte management [1]. The physiochemical properties, water uptake, membrane performance, and short-term and long-term behaviour of PBI/H_3PO_4 membranes was studied by many groups [9-11,42]. Schmidt [43] summarized the durability and degradation of a Celtec®-P 1000 MEA. The properties of this product running in start/stop operation mode were discussed in [44]. It should not be operated below 100°C because below this point, liquid water exists within the cell, causing an irreversible loss of phosphoric acid. For an operating temperature above 100°C, the water within the HTPEM fuel cell is gaseous, eliminating the risk of porous media flooding and minimizing water management issues. At a typical operating temperature above 130-140°C, the conductivity of phosphoric acid under anhydrous conditions decreases due to the formation of pyrophosphoric acid ($H_4P_2O_7$) by the condensation of two molecules of phosphoric acid and the elimination of water (dehydration). Thus, the cell resistance of the PBI/H_3PO_4 membrane is higher at 160°C. Under electrical load, water is produced by the fuel cell reaction, rehydrates the membrane and shifts the equilibrium between phosphoric acid and pyrophosphoric acid towards the more conductive phosphoric acid [1,45]. With a high operating temperature it is possible to design smaller cooling systems due to the large temperature difference compared to the surroundings. High operating temperatures result in faster electrode kinetics and a CO tolerance that is higher by several orders of magnitude compared to LTPEM fuel cells. For the membrane electrode assembly, carbon supported platinum is needed for effective catalysis of the anode, and the cathode side reactions. At the cathode side, platinum alloy catalysts are used to maximize the cell performance. Polytetrafluoroethylene (PTFE) is often used as a binder and hydrophobic agent with a weight fraction of 20–60% [1]. The CO tolerance is directly related to the thermodynamics of CO and hydrogen on the platinum catalyst. In [25], the authors demonstrated that with 1% CO in the fuel gas, a HTPEM fuel cell exhibits poor performance below 120°C, whereas at a temperature of 175°C, even more than 10% CO can be used without significantly affecting the cell performance [1,42]. Thus, it is possible to directly feed reformed gas containing a CO content of several percent into the HTPEM fuel cell. CO poisoning of platinum is reversible, and the cell performance is restored after switching from reformate to pure hydrogen.

2.5 Future development

Although major advances have been achieved in HTPEM fuel cell technology in recent years, some problems still persist. The durability of such systems is acceptable under steady-state operation conditions within the laboratory, whereas temperature and load cycling influence the system durability. Under defined steady-state operating conditions, a degradation rate of less than 6 µV h^{-1} has been obtained over a period of 20,000 h [38]. In [46], the authors stated that acid loss is not an important cause of degradation under appropriate operating conditions. Instead, under steady-state operating conditions, the loss of active platinum area, carbon support corrosion, and increased transport resistance are the primary reasons for performance loss [1]. Thus, increasing the performance of the HTPEM fuel cell is an important issue. The performance could potentially be improved by decreasing the phosphate anion adsorption, thus increasing the oxygen solubility and diffusivity. Because phosphoric acid is located not only within the membrane but also within the reaction layer, efficient methods for fixing the acid at the appropriated location must be developed (see, e.g., [47-50]). With respect to the mechanical long-term stability, researchers must systematically address the stress-relaxation behaviour of membranes under dynamic conditions. Further research is needed to reduce the noble metal loadings and to investigate for alternative bipolar-plate material (metallic bipolar-plates or thin silicon plates) [1]. Especially for HTPEM fuel cell stacks, it is important to establish a homogeneous temperature distribution. Current issues are directly related to temperature cycling, tempering concepts, and heat-up and shut-down procedures (see, e.g., [51-55]). Most of the above discussed topics are related to the layout of the key components and an understanding of the internal quantities distributions. In terms of HTPEM fuel cell modeling and simulation, further work is needed to completely understand the electrolyte behaviour.

2.6 Fuel cell thermodynamics

2.6.1 Potential of a fuel cell

Thermodynamics is the study of the transformation of energy from one form to another. Because fuel cells are energy conversion devices, fuel cell thermodynamics is the key to understand the conversion of chemical energy into electrical energy. Moreover, thermodynamic studies can place upper bound limits on the maximum electrical potential that can be generated and can explain the theoretical boundaries of what is possible with a fuel cell. The following equations briefly summarize fuel cell thermodynamics based on the textbooks [17-20]. For an electrochemical system, such as a fuel cell, the Gibbs free energy provides the maximum amount of energy available to do electrical work. Irreversible losses in energy conversion due to the creation of entropy cannot be converted into useful electrical work (Eq.(4)).

$$G = U - T \cdot \Delta S + p \cdot \Delta V \tag{4}$$

A change in the Gibbs free energy is given by Eq.(5).

$$\Delta G = \Delta U - T \cdot \Delta S - S \cdot \Delta T + p \cdot \Delta V + V \cdot \Delta p \tag{5}$$

Based on the first law of thermodynamics, the work term can be expanded to include both mechanical and electrical work and Eq.(5) can be rewritten. It can be reasonably assumed that fuel cells operate at a constant pressure and temperature during the reaction process. The maximum achievable electrical work for an electrochemical system under isobaric and isothermal conditions is defined by the change in the Gibbs free energy ($\Delta p = 0$, $\Delta T = 0$).

$$\Delta U = T \cdot \Delta S - \Delta W$$
$$\Delta U = T \cdot \Delta S - \left(p \cdot \Delta V + \Delta W_{el} \right)$$
$$\Delta G = \left(T \cdot \Delta S - \left(p \cdot \Delta V + \Delta W_{el} \right) \right) - T \cdot \Delta S - S \cdot \Delta T + p \cdot \Delta V + V \cdot \Delta p \tag{6}$$
$$\Delta G = -\Delta W_{el}$$
$$W_{el} = -\Delta G$$

The potential of a system to perform electrical work is measured by the voltage, also known as the electrical potential. The electrical work performed by moving a charge through an electrical potential difference is given by Eq.(7).

$$W_{el} = E \cdot Q \tag{7}$$

If the charge is assumed to be carried by electrons, Eq.(7) can be rewritten as follows:

$$Q = n \cdot F$$
$$W_{el} = E \cdot n \cdot F \tag{8}$$

The theoretical maximum cell potential for a hydrogen-oxygen fuel cell at standard temperature and pressure (STP) is calculated from Eq.(9), returning a value of 1.229 V. If gaseous product water is considered, this value decreases to 1.18 V.

$$E^0 = -\frac{\Delta G}{n \cdot F} \tag{9}$$

Fuel cells are normally operated at non-standard conditions. Thus, the reversible fuel cell potential is affected by temperature, pressure, and species activity or concentration. The variation in reversible potential with temperature is given by Eq.(10) and Eq.(11).

$$\left(\frac{\Delta G}{\Delta T}\right)_p = -S \tag{10}$$

Along with the previous equations, the reversible cell potential varies with temperature, as expressed by Eq.(11).

$$\left(\frac{\Delta E}{\Delta T}\right)_p = \frac{\Delta S}{n \cdot F} \tag{11}$$

The reversible cell potential at an arbitrary temperature and constant pressure is calculated with Eq.(12).

$$E_T = E^0 + \frac{\Delta S}{n \cdot F} \cdot (T - T_0) \tag{12}$$

The theoretical cell potential decreases with increasing temperature. However, when a fuel cell is operating, a higher cell temperature generally results in a higher potential because the individual potential losses decrease with increasing temperature and more than compensate for the losses of the theoretical cell potential [17].

Similar to the temperature effects, the effects of pressure on the cell potential are calculated as follows:

$$\left(\frac{\Delta G}{\Delta p}\right)_T = V$$
$$\left(\frac{\Delta E}{\Delta p}\right)_T = -\frac{\Delta V}{n \cdot F} \tag{13}$$

The variation of the reversible cell potential with pressure is related to the volume change of the reaction. If the volume change of the reaction is negative, the cell potential increases with increasing pressure. The reversible potential also varies with concentration. The chemical potential measures how the Gibbs free energy of a system changes with variations in the chemistry of a system. The chemical potential is related to concentration through the activity.

$$\mu_i = \mu_i^0 + R \cdot T \cdot \ln(a_i) \tag{14}$$

The Gibbs free energy can be calculated for a system of i chemical species using Eq.(15).

$$\Delta G = \sum_i \left(\mu_i^0 + R \cdot T \cdot \ln(a_i) \right) \cdot \Delta n_i \tag{15}$$

Along with the above equations, the reversible cell potential can be expressed for a system with an arbitrary number of products and reactant species as follows.

$$E = E^0 - \frac{R \cdot T}{n \cdot F} \cdot \ln \frac{\prod a_{prod}^{v_i}}{\prod a_{react}^{v_i}} \tag{16}$$

For a hydrogen-oxygen fuel cell operating below 100°C, the product water is in its liquid form and the Nernst equation is given by Eq.(17).

$$E = E^0 - \frac{R \cdot T}{n \cdot F} \cdot \ln \frac{a_{H_2O}}{a_{H_2} \cdot a_{O_2}^{0.5}} \tag{17}$$

Replacing the activities with partial pressures, setting the activity of water to unity, and introducing the reversible cell potential at an arbitrary temperature leads to Eq.(18).

$$E = \left(E^0 + \frac{\Delta S}{n \cdot F} \cdot (T - T_0) \right) - \frac{R \cdot T}{n \cdot F} \cdot \ln \frac{1}{p_{H_2} \cdot p_{O_2}^{0.5}} \tag{18}$$

The above equation it is also seen, that the potential decreases once the reactants are diluted, for example if air is used instead of pure oxygen.

2.6.2 Efficiency of a fuel cell

The efficiency of an energy conversion device is defined as the ratio between useful energy output and energy input. The energy input is the enthalpy of the hydrogen, which is its higher heating value [17]. The theoretical maximum efficiency of a fuel cell is calculated with Eq.(19), which assumes that all of the Gibbs free energy is converted into electrical energy, resulting in an efficiency of 83%.

$$\eta_0 = \frac{\Delta G}{\Delta H} \qquad (19)$$

Eq.(19) illustrates that the fuel cell efficiency is proportional to the cell operating potential. The real fuel cell efficiency during operation is significantly below the ideal fuel cell efficiency, primarily due to potential losses and fuel utilization losses.

$$\eta = \eta_0 \cdot \eta_{voltage} \cdot \eta_{fuel} \qquad (20)$$

The potential efficiency accounts for several fuel cell losses which can be captured by the current-voltage-curve (I-V-curve), expressed as the ratio between the operating potential and reversible potential. The fuel utilization losses account for the fact that not all of the fuel provided participates in the electrochemical reactions. may undergo side reactions that do not produce electrical power, may flow through the fuel cell without ever reacting, or may permeate through the membrane. The ratio of the fuel used by the cell to generate electrical power versus the total fuel provided to the cell is given by Eq.(21).

$$\eta_{fuel} = \frac{I}{n \cdot F} \cdot \frac{1}{v_{fuel}} \qquad (21)$$

For fuel cells operating at stoichiometric conditions, the fuel utilization is independent of the current and can be rewritten, as follows:

$$\eta_{fuel} = \frac{1}{\lambda} \qquad (22)$$

Introducing the above equations into Eq.(20) finally leads to Eq.(23) which expresses the efficiency for a fuel cell operating at stoichiometric flow rates.

$$\eta = \frac{\Delta G}{\Delta H} \cdot \frac{E_{cell}}{E} \cdot \frac{1}{\lambda} \qquad (23)$$

2.7 Fuel cell characterization methods

2.7.1 Current-voltage-curve (I-V-curve) of a PEM fuel cell

The typical I-V-curve of a PEM fuel cell is characterized by several potential losses occurring at no-load and load operating conditions (Fig.2). The curve represents the potential output of the fuel cell for a given load current and is typically recorded by starting at an open circuit potential and then by increasing the current and taking measurements at prescribed current intervals [17]. The potential measured during operation is always lower than the theoretical maximum cell potential. As more current is drawn from the cell, the lower the potential output of the cell, limiting the total power that can be delivered.

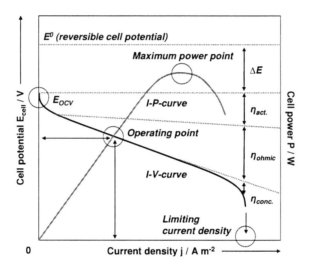

Fig.2. Fuel cell I-V-curve and potential losses during operation of a PEM fuel cell.

According to [17,20], the I-V-curve can be divided into three regions. At a low current density, activation losses primarily affect the initial part of the curve. These losses are associated with sluggish electrode kinetics, processes involving the adsorption of reactant species, the desorption of product species, the transfer of electrons across the double layer, and the nature and type of the surface involved. These losses occur at both sides of the cell. For an electrochemical system, the exponential relationship between the current density and activation overpotential is described by the simplified Butler-Volmer equation.

$$j = j_0 \cdot \left(e^{\frac{\alpha \cdot n \cdot F \cdot \eta_{act}}{R \cdot T}} - e^{\frac{-(1-\alpha) n \cdot F \cdot \eta_{act}}{R \cdot T}} \right)$$

(24)

The activation overpotential represents the voltage (loss) needed to overcome the activation barrier associated with the electrochemical reactions. When the activation overpotential is large, the second term in Eq.(24) becomes negligible and the Butler-Volmer equation takes the following form:

$$j = j_0 \cdot \left(e^{\frac{\alpha \cdot n \cdot F \cdot \eta_{act}}{R \cdot T}} \right) \tag{25}$$

Solving this equation for the activation overpotential leads to Eq.(26), also known as the generalized Tafel equation.

$$\eta_{act} = -\frac{R \cdot T}{\alpha \cdot n \cdot F} \cdot \ln j_0 + \frac{R \cdot T}{\alpha \cdot n \cdot F} \cdot \ln j \tag{26}$$

In the intermediate region, ohmic losses dominate the shape of the I-V-curve. These losses occur due to electronic and ionic conduction, the resistance of all the materials (e.g., the copper current collector, bipolar-plate, gas diffusion layer, microporous layer, reaction layer, and membrane), and all contact resistances.

$$\eta_{ohmic} = I \cdot \left(R_{el} + R_{ionic} \right) \tag{27}$$

At a high current density, concentration losses occur due to mass transport. Reactants are rapidly consumed at the respective electrodes by the ongoing electrochemical reactions leading to a concentration gradient is present. Species consumption and accumulation in the reaction layer lead to a performance loss. This performance loss includes the Nernst loss and reaction rate loss and is given by the following equation:

$$\eta_{conc} = \frac{R \cdot T}{n \cdot F} \cdot \left(1 + \frac{1}{\alpha} \right) \cdot \ln \frac{j_L}{j_L - j} \tag{28}$$

Additional losses might occur due to fuel crossover, namely, the loss due to the direct electrochemical reaction of permeated oxygen at the anode and of hydrogen at the cathode [20].

2.7.2 Electrochemical impedance spectroscopy

The I-V-curves represent the actual performance of a fuel cell at given operating conditions but do not provide detailed information about the ongoing mechanisms that define the cell performance. EIS allows the investigation of the impedance in electrochemical power generation and basic electrochemical research [56]. EIS is a powerful method for identifying and distinguishing between the mechanisms within a fuel cell such as the anode side reaction, the cathode side reaction, transport processes, or high frequency resistance. Depending on the frequency, the impedance of the fuel cell will change depending on the time scale of the ongoing mechanisms. EIS measurements are performed by superimposing a small sinusoidal voltage perturbation (potentiostatic mode) or current perturbation (galvanostatic mode) over the fuel cell to record its response at different frequencies. The excitation signal may have the following form.

$$I = I_0 \cdot \sin(\omega \cdot t) \tag{29}$$

The relationship between the radial frequency and frequency is given by Eq.(30).

$$\omega = 2 \cdot \pi \cdot f \tag{30}$$

The response signal of the system is shifted in phase and may have the following form:

$$U = U_0 \cdot \sin(\omega \cdot t - \varphi) \tag{31}$$

According to Ohm's law, the impedance of the system can be calculated from Eq.(32).

$$Z = \frac{U}{I} = \frac{U_0 \cdot \sin(\omega \cdot t - \varphi)}{I_0 \cdot \sin(\omega \cdot t)} = Z_0 \cdot \frac{\sin(\omega \cdot t - \varphi)}{\sin(\omega \cdot t)} \tag{32}$$

In complex notation, the impedance of a fuel cell can be calculated with the following equation.

$$Z = Z_0 \cdot e^{j \cdot \varphi} = Z_0 \cdot (\cos\varphi + j \cdot \sin\varphi) \tag{33}$$

According to Eq.(33), the impedance of a system can be expressed in terms of a real part and an imaginary part.

$$Z_{RE} = Z_0 \cdot \cos(\varphi)$$
$$Z_{IM} = Z_0 \cdot \sin(\varphi \cdot j)$$
$$(34)$$

The recorded EIS data is generally represented as a Bode or Nyquist plot. The Bode plots contain detailed frequency information. The impedance values and phase of a system are plotted against the logarithmic frequency. In Nyquist plots, the impedance is given as a vector of length Z and the angle between this vector and the real part axis (Fig.3(b)).

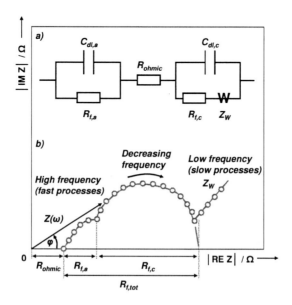

Fig.3. Typical equivalent circuit model (a) and Nyquist plot (b) of a fuel cell (similar to that shown in [20]).

EIS can be used to characterize and represent the behaviour of a fuel cell in terms of an equivalent circuit model. This means that the individual contributions of the physical components and processes can often be identified and determined due to their individual frequency dependencies and different positions in the circuit (Fig.3(a)). Equivalent circuit models can be assembled from a set of basic impedance elements that are connected in a specific way to build equivalent circuits, which should represent the electrical behaviour of the tested system under the given operating conditions [56]. A simple equivalent circuit model of a fuel cell is described using ohmic resistors, parallel resistance capacitor elements (RC elements), and a Warburg element [20]. Other typical elements are the constant phase element or the advanced tools necessary for modeling porous electrodes. The ohmic resistor is used to describe the ohmic losses. Its value is given by the high frequency (first) real axis intercept in the Nyquist plot. The two RC elements are used to model the anode and cathode activation losses. These anode activation losses are represented in the Nyquist plot by the diameter of the first (high frequency) loop, and the cathode activation losses are indicated by the (generally larger) diameter of the second loop. Additionally, an infinite Warburg element can be used to describe cathode mass transport effects [20]. According to [26], such a schematic representation of a Nyquist plot is easy to understand but may not be accurate. In most cases, the loops in the high frequency range and low frequency ranges cannot be separated in this manner because events at the anode side may influence the cathodic loss region and vice versa. More information on EIS or cyclic voltammetry can be found in [57-59]. A discussion on the artifacts that may appear at the lowest and highest frequencies and the contributions of the electrode impedance can be found in [56].

3. PEM fuel cell modeling

Because of the spatial dimensions of a fuel cell, it is difficult to measure internal quantities distributions. This fact has prompted researchers to develop various models to understand the operational behaviour of fuel cells. The use of mathematical models is one possibility for analyzing species concentrations, temperature gradients, and pressure distributions in each component of the fuel cell. Thus, it becomes possible to predict the internal workings of a fuel cell, for different operating conditions and layouts. Reviews of fuel cell modeling and simulation have been published by Cheddie and Munroe [60], Haraldsson and Wipke [61], Djilali [62], Weber and Newman [63], Wang [64], Yao et al. [65], Bıyıkoğlu [66], Faghri and Guo [67], Costamagna and Srinivasan [68,69], Siegel [70], and Mench [1]. These publications discuss fuel cell related micro-, and macroscopic governing equations, modeling strategies, modeling approaches, computational domains, complexity and details, and common modeling assumptions. Some of the reviews elucidate the validation of such models and future research needs. According to [63], a fuel cell model can be represented by the following five types of relations: (1) conservation laws of mass, momentum, energy, species, and charge; (2) the constitutive relationships for various fluxes; (3) kinetic equations for reactions; (4) equilibrium relationships; and (5) auxiliary and supporting relationships, including variable definitions. All relationships are strongly coupled to each other in addition to the material properties, empirical relationships, and experimental data. Conservation laws can be written using the following partial differential equation (PDE) [71].

$$c_1 \cdot \frac{\partial^2 \xi}{\partial t^2} + c_2 \cdot \frac{\partial \xi}{\partial t} + \nabla \cdot \left(-c_3 \cdot \nabla \xi - c_4 \cdot \xi + S_1\right) + c_5 \cdot \nabla \xi + c_6 \cdot \xi = S_2 \qquad (35)$$

In Eq.(35), ξ is the general variable to be solved for, c_1 is a mass coefficient, c_2 is a damping mass coefficient, c_3 and c_4 are diffusion and convective coefficients (conservative flux), c_5 is a convective coefficient, and c_6 is an absorption coefficient. S_1 is a source or sink term of the conservative flux term, and S_2 is a general source or sink term. The boundary conditions (generalized Neumann and Dirichlet) are summarized by Eq.(36) [71], c_7 and c_8 are boundary coefficients, S_3 and S_4 are boundary source terms, and ψ is a Lagrange multiplier (defined at the boundary).

$$\vec{n} \cdot \left(c_3 \cdot \nabla \xi + c_4 \cdot \xi - S_1\right) + c_7 \cdot \xi = S_3 - c_8^T \cdot \psi$$
$$c_8 \cdot \xi = S_4 \qquad (36)$$

The state of a model differs between steady-state and dynamic (transient). The latter depends on the time derivatives in Eq.(35). The time constants for the electrochemical double layer treatment, overall heat and mass transfer, and membrane hydration and dehydration vary by several orders of magnitude. Estimations of time constants can be found in [63,64]. Dynamic models are often used when analyzing step changes in operating conditions or when analyzing start-up and shut-down procedures.

3.1 Overall modeling aspects

Depending on the needs of a researcher, either a system model (analytic, semi-empirical, or empirical) or mechanistic model can be used. Theoretical models (mechanistic at the macroscopic or microscopic level) are more complex and include minute details of the fuel cell operation. When localized phenomena at the pore level size or at a single three-phase boundary must be investigated, detailed microstructural knowledge should be incorporated. Depending on the simulative application, computational efforts may be limited to single fuel cell components. Complete fuel cells models can be computed but require more effort and time to solve. Currently, complete fuel cell stacks including auxiliary components are mainly simulated using system models. For models that are solved using finite differences, finite volumes or finite elements, the computational domain must be carefully selected. When using a single-domain, only the source or sink terms will vary according to the position within the cell (no internal boundary conditions). All equations are written in the form of a generic convection-diffusion equation, and all terms that do not fit that format are combined in the source or sink term [60]. Multi-domain models use different equations in each modeling domain and require a careful handling of the boundary, initial, internal boundary (e.g., continuity), and external boundary condition (e.g., exact operating parameters that are measured and controlled in fuel cell applications). Decades ago, researchers used one-dimensional models with various degrees of complexity. Fluxes, concentrations, temperatures, and cell voltages were analyzed for given boundary conditions and were taken to be normal to the computational subdomains. Two-dimensional models may represent an improvement over one-dimensional models. These models offer a more realistic view of certain phenomena because spatial distributions are considered. These models may use a sandwich domain (y-z-plane) or an along-the-channel domain (x-z-plane). Sandwich models are primarily used for the analysis of fluxes, heat and mass transfer and concentrations, including the effect of the bipolar-plate and gas channel. Along-the-channel model domains can be used to analyze quantities along the gas channel. One-, and two-dimensional models may include the same conservation equations as three-dimensional models, and thus, they provide considerable information with sufficient accuracy if the boundary and initial conditions are carefully selected. A three-dimensional model is most appropriate for analyzing the overall fuel cell behaviour. Such a model can be seen as a combination of both two-dimensional computational domains (y-z-plane and x-z-plane) and has the ability to study the blocking effect of the bipolar-plates, the detailed current density distribution, or the effectiveness of a flow-field design. With the currently available computational power, simulations are becoming increasingly complex and include a high level of details such as two-phase flow, liquid water formation, and water transport within the channels. Many of the simplifications and assumptions involved are directly linked to the modeling and simulation investigation itself. The use of simplifications does not necessarily imply that the modeling results are wrong. Every model is as good as the assumptions on which it is built, and it is important to understand these assumptions to predict the model's limitations and accurately interpret its results [17]. Additional theoretical background on fuel cell modeling can be found in, e.g., [63,64,72].

3.2 HTPEM fuel cell models – Literature review

As reported in previous reviews [60-70], LTPEM fuel cell models exist for over two decades, whereas only a handful of publications concerning HTPEM fuel cell modeling and simulation are currently available in the literature. Korsgaard et al. [73] developed a simple semi-empirical model to describe their experimental data, and good agreement was found. Cheddie at al. [74-77] published one-, two-, and three-dimensional models accounting for different operating conditions, membrane and reaction layer properties, layout optimization, gas solubility, and gaseous dissolution into the aqueous phase. Steady-state and transient three-dimensional HTPEM fuel cell models were presented by Peng et al. [78,79]. The authors demonstrated that thermal management strongly affects the fuel cell performance, and discussed key optimization parameters for performance improvements. Scott et al. [80] proposed a one-dimensional model that could satisfactorily predict the performance curve. This model was used to simulate the effects of catalyst loading and the Pt/C-ratio on fuel cell performance. Ubong et al. [81] developed a single-channel three-dimensional model in which the reaction layer was assumed to be infinitely thin and the electrochemical reactions were described using an agglomerate approach. A complete three-dimensional model was developed and solved in [82], highlighting reaction layer kinetics. Shamardina et al. [83] presented a simple and quickly solvable steady-state, isothermal, pseudo two-dimensional model that accounted for crossover effects. Another analytical HTPEM fuel cell model, published by Kulikovsky et al. [84], included important basic kinetic and transport parameters. A two-dimensional isothermal model was published by Sousa et al. [85], who treated the reaction layer as spherical catalyst agglomerates with porous interagglomerate spaces. The model was used to study the influence of the reaction layer properties on cell performance. A control-oriented, one-dimensional model was developed in [86], addressing the transient responses of a HTPEM fuel cell. Wang et al. [87] investigated the transient evolution of the CO poisoning effect of PBI membrane fuel cells using a one-dimensional model. Another work that deals with the CO poisoning and its dynamics was presented in [88]. Various HTPEM fuel cell models were presented in [89-92], primarily addressing the fluid-(gas)-, and the solid-phase temperature and the PBI/H_3PO_4 sol-gel membrane behaviour. Structural mechanics, namely, localized fluid-structural interactions, were analyzed in [92]. In [93], a complete HTPEM fuel cell was modeled and the predicted values were compared to measurements for typical operating conditions. The simulated current density distribution and the fluid-(gas)-, and the solid-phase temperature distributions were also analyzed. The Arrhenius approach is found to be valid within a defined temperature range and may overpredict the PBI/H_3PO_4 sol-gel membrane conductivity at a higher solid-phase temperature. Moreover, the influence of the fluid-(gas)-phase temperature on the solid-phase temperature was investigated. Kvesić et al. [94] published a three-dimensional model of a 200 cm^2 HTPEM five cell short stack. The model was presented in the form of a multi-domain, multi-scale model, that allowed for the simulation of an entire stack with reasonable computational power and time. Simulation results were compared to segmented measurements of temperature and current density. Lüke et al. [95] presented a performance analysis of a HTPEM fuel cell stack, addressing temperature and current density measurements acquired while using pure hydrogen and synthetic reformate as the anode gas. A three-dimensional model was used for validation purposes. The authors found that oxygen depletion was the primary cause of the uneven current density distribution. Moreover, it was shown that homogenization can be achieved without a reduction of the stack voltage if the stack is operated with reformate when switching from co-flow to counter-flow configuration. Two three-dimensional models (an agglomerate model and pseudohomogenous model) were presented in [96], focusing on different approaches for modeling HTPEM fuel cells. These two modeling approaches were discussed, and the experimental results demonstrated that the pseudohomogeneous model produced a better fit. Sousa et al. [97] presented a nonisothermal model of a HTPEM fuel cell, treating the catalyst layers as spherical catalyst particle agglomerates with a porous interagglomerate space. The authors investigated the influence of different modeling geometries on the performance and found that the along-the-channel model did not represent the general performance trend, concluding that this particular modeling geometry is not appropriate for fuel cell simulations. The same group published a one-dimensional model for analyzing the influence of CO, carbon dioxide, and methane, which would be present in a reformate gas, considered in terms of the effect at the anode polarisation and kinetics behaviour [98]. Kurz presented an interesting method for modeling and simulation of a HTPEM fuel cell using a coupled two-, and three-dimensional approach [99]. In [100], a parametric study of the external coolant system of a HTPEM fuel cell was published, focusing on the temperature variations within the stack, the number of coolant plates, and the coolant flow. In [101], the effect of CO poisoning in HTPEM fuel cells with various types of flow-fields was numerically studied.

4. The developed HTPEM fuel cell model

4.1 Model geometry

The three-dimensional model geometry used for all modeling and simulation is shown in Fig.4 and Fig.5. This model can represent the fluid-flow distribution within the gas channels and the internal quantities distributions over the membrane electrode assembly area. The model is an exact, nearly one-to-one representation of the HTPEM fuel cell that was used for all experimental tests and includes the following computational subdomains: The gold-plated copper current collectors (CU), the high temperature stable bipolar-plates, the flow-fields, and the membrane electrode assembly that is sandwiched between the two bipolar-plates. The membrane electrode assembly includes the gas diffusion layers, the reaction layers, and the PBI/H_3PO_4 sol-gel membrane.

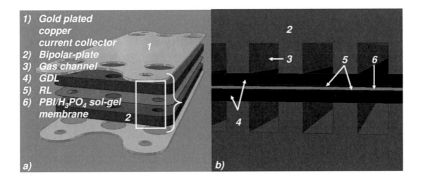

Fig.4. Computer aided design (CAD) representation of the HTPEM fuel cell (x-y-z-plane) (a). Detailed view of the involved components (x-y-z-plane) – compare with Fig.1 (b).

Table 2 summarizes the geometrical aspects of the different components of the HTPEM fuel cell model used for all simulations. The reported values were obtained from optical measurements collected at room temperature and pressure, as shown in Fig.1. Although various flow-fields of different shapes and sizes have been investigated over the past decades in fuel cell research, the development of a particular design is still a complex topic because the milled flow-fields must supply the gases to the reaction layer and must remove the water produced. Moreover, the design must ensure good electrical and thermal contact between the components. An optimized flow-field will lead to a more homogeneous temperature and current density distribution. In this work, three types of flow-field subdomains were incorporated using the same model geometry shown in Fig.5. The three types of flow-fields are designed such that the positioning of the inlets and outlets remains unchanged, as determined by the inhouse design. The black and red arrows in Fig.5 indicate the gas flow direction in the co-flow and counter-flow configuration.

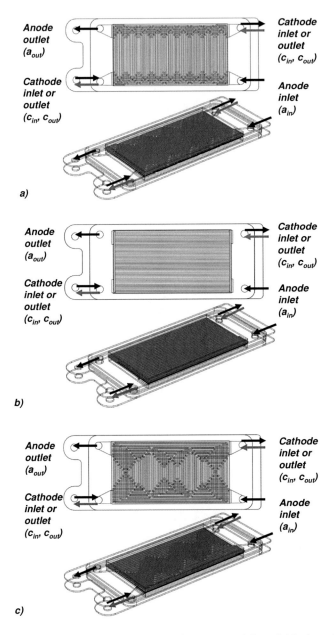

Fig.5. 3D model of the HTPEM fuel cell with the three types of flow-fields (x-y-plane and x-y-z-plane) – 6 channel parallel serpentine (a), 26 channel parallel straight (b), and mixed-type (c).

Table 2

Geometrical aspects of the different components of the HTPEM fuel cell model.

Subdomain	Parameter	Value	Note
		mm	
CU	l_{CU}	1.000	thickness (measured)
BPP	l_{BPP}	3.000	thickness (measured)
GDL	$l_{0,GDL,a}$	0.400	thickness uncompressed (measured)
	$l_{0,GDL,c}$	0.425	thickness uncompressed (measured)
RL	$l_{RL,a}$	0.030	assumed
	$l_{RL,c}$	0.040	assumed
MEM	l_{MEM}	0.152	measured

4.1.1 Type I flow-field

Type I flow-field is a 6 channel parallel serpentine configuration (Fig.5(a)). A good availability of the gases over the membrane electrode assembly area is assumed because the gases are expected to strictly follow the gas channel. The main drawback of this flow-field is the relatively large pressure drop between the inlets and outlets. Additionally, the gases are continuously consumed towards the outlet, resulting in lower current density values towards the air outlet. All 6 channels are in direct contact with the gas inlet and outlet.

4.1.2 Type II flow-field

Type II flow-field is a 26 channel parallel straight configuration (Fig.5(b)). All channels are connected by one lateral gas channel. The 5 outermost channels, located at the left and right side of the cell, are in direct contact with the gas inlet and outlet. The main advantage of this configuration is the low pressure drop, whereas a poor distribution of the fluid-flow is expected in the middlemost gas channel region.

4.1.3 Type III flow-field

Type III flow-field is a combination of the serpentine-parallel-, and parallel-type configurations (Fig.5(c)). The entire flow-field is divided into three regions so that these regions can be fed with oxygen rich air. Similar to type I flow-field, 6 gas channels are in direct contact with the gas inlet and outlet, whereas 2 channels supply gas to the lower part, 2 channels to the middle part, and 2 channels to the upper part. The advantages of this configuration includes its relatively low pressure drop and the fact that high oxygen concentrations are available in different regions over the membrane electrode assembly area.

Table 3

Geometrical aspects of the three types of flow-fields.

Type	Channel volume	Channel to wall area	Channel/land ratio	Depth	Mean channel length
	m^3	m^2	mm / mm	mm	mm
I	$2.7348 \cdot 10^{-6}$	0.008124	1 / 1	1	450
II	$2.6345 \cdot 10^{-6}$	0.007734	1 / 1	1	50 (lateral channels)
					100 (straight channels)
III	$2.5641 \cdot 10^{-6}$	0.007627	1 / 1	1	225

Table 3 lists some geometrical aspects of the three types of flow-field geometries under study. This table illustrates that type I flow-field has the highest channel volume and highest channel to wall area for heat transfer between the fluid-(gas)-, and the solid-phase, followed by type II flow-field and type III flow-field. Nevertheless, the reported values are in the same order of magnitude.

4.2 Subdomain transport equations

4.2.1 Momentum transport

The continuity equation and incompressible Navier-Stokes equations (where the density is assumed to be constant or nearly constant) are solved to account for the laminar gas flow and pressure distribution within the gas channels. Eq.(37) accounts for the advection momentum flux and momentum imparted due to pressure and viscosity.

$$\nabla \cdot u = 0$$
$$\rho \cdot (u \cdot \nabla) \cdot u = \nabla \cdot \left(-p \cdot I + \eta \cdot \left(\nabla u + (\nabla u)^T \right) \right) \tag{37}$$

The Brinkman equations describe gas flow within porous media. This mathematical model extends Darcy's law to include a term that accounts for viscous transport in the momentum balance, and it treats both the pressure and the flow-velocity vector as independent variables.

$$\nabla \cdot u = \frac{S_m}{\rho}$$

$$\left(\frac{\eta}{k_p} + S_m \right) \cdot u = \nabla \cdot \left(-p \cdot I + \frac{1}{\varepsilon} \cdot \left(\eta \cdot \left(\nabla u + (\nabla u)^T \right) - \left(\frac{2}{3} \cdot \eta \right) \cdot (\nabla \cdot u) \cdot I \right) \right) \tag{38}$$

4.2.2 Mass (species) transport

The gas flow is predominantly convective in the gas channels, whereas it is predominantly diffusive within the porous media. The mass (species) transport is solved using Stefan-Maxwell diffusion and the convection application mode while accounting for hydrogen and water at the anode side and oxygen, water, and nitrogen at the cathode side (Eq.(39)). The reaction rates appear as source/sink terms.

$$\nabla \cdot \left(\rho \cdot \omega_i \cdot u - \rho \cdot \omega_i \cdot \sum_{j=1}^{n} \tilde{D}_{ij}^{eff} \cdot \left(\nabla x_j + (x_j - \omega_j) \cdot \frac{\nabla p}{p} \right) \right) = S_{\omega_i} \tag{39}$$

The mass transport source/sink terms are non-zero only in the reaction layer subdomains. The reaction rates for the different species are described by Eq.(40).

$$S_{\omega_i} = \begin{cases} -j_a \cdot \dfrac{M_{H_2}}{2 \cdot F} \\ j_c \cdot \dfrac{M_{O_2}}{4 \cdot F} \\ -j_c \cdot \dfrac{M_{H_2O}}{2 \cdot F} \end{cases} \tag{40}$$

4.2.3 Energy transport

The thermal behaviour is described using a two-equation system that represents the thermal interactions between the two phases due to the large temperature difference between the gas temperature (fluid-(gas)-phase temperature) and the cell operating temperature (solid-phase temperature) [89,90]. The solid-phase temperature distribution is calculated for the gold-plated copper current collectors, the bipolar-plates, the solid matrix of the porous media, and the membrane. The fluid-(gas)-phase temperature distribution is solved within the gas channel and within the porous media. Both equations are coupled through their source/sink terms and account for possible heat transfer between the fluid-(gas)-, and the solid-phase using a volumetric heat transfer coefficient. In general, this coefficient is a function of the morphology of the porous media. As stated in [102], typical values for metal foams vary from $2 \cdot 10^4$ to $2 \cdot 10^5$ W m^{-3} K^{-1} for porosities between 0.7 and 0.95.

$$\nabla \cdot \left(-k \cdot \nabla T_s \right) = S_{T_s}$$
$$\nabla \cdot \left(-k \cdot \nabla T_f \right) = S_{T_f} - \rho \cdot C_p \cdot u \cdot \nabla T_f \tag{41}$$

The source/sink terms for the solid-phase temperature are given in Eq.(42) and account for ohmic and protonic heating (membrane heating), irreversible reaction heat, and reaction entropy for the following subdomains: the gold-plated copper current collector, the bipolar-plates, the gas diffusion layer, both reaction layers, and the membrane (top to bottom).

$$S_{T_s} = \begin{cases} \dfrac{i_s^2}{\sigma_s} \\[2mm] \dfrac{i_s^2}{\sigma_s} - h_{T_s/T_f} \cdot \left(T_s - T_f \right) \\[2mm] \dfrac{i_s^2}{\sigma_s} + \dfrac{i_m^2}{\sigma_m} + j_a \cdot \eta_a - h_{T_s/T_f} \cdot \left(T_s - T_f \right) \\[2mm] \dfrac{i_s^2}{\sigma_s} + \dfrac{i_m^2}{\sigma_m} + j_c \cdot \eta_c + j_c \cdot \dfrac{\Delta S_c \cdot T_s}{4 \cdot F} - h_{T_s/T_f} \cdot \left(T_s - T_f \right) \\[2mm] \dfrac{i_m^2}{\sigma_m} \end{cases} \tag{42}$$

Eq.(43) introduces the source/sink term for the fluid-(gas)-phase temperature within the porous media.

$$S_{T_f} = h_{T_s/T_f} \cdot \left(T_s - T_f \right) \tag{43}$$

4.2.4 Charge transport

Two Poisson equations are used to evaluate the charge transport. The solid-phase potential is solved within the gold-plated copper current collectors, the bipolar-plates, and the porous media. The membrane-phase potential is solved within the reaction layer and in the membrane. Both equations are coupled through the current source/sink term.

$$-\nabla \cdot \left(\sigma_s \cdot \nabla \phi_s \right) = -S_\phi$$
$$-\nabla \cdot \left(\sigma_m \cdot \nabla \phi_m \right) = +S_\phi \tag{44}$$

4.2.5 Electrochemistry and transport properties for the spherical agglomerate model

Agglomerate models offer a physical and mathematical description of the transport processes occurring in the reaction layer based on analytical solutions to coupled species diffusion and electrochemical half-cell reactions [103-105]. These models were developed to describe the behaviour of various types of fuel cells, including PAFCs, LTPEM fuel cells, and HTPEM fuel cells [97,98,106-108]. The agglomerate structure model used in this work is based on the LTPEM fuel cell model from the COMSOL Multiphysics® model library [71] and on the equations presented in [107,109]. The reader is referred to [110] for a detailed description of the analytical solution of a diffusion reaction problem for a spherical porous particle. The reaction layer is assumed to consist of numerous spherical agglomerates (Fig.6). The spherical agglomerates consist of a carbon black support with platinum on the surface. The zones filled with amorphous phase phosphoric acid are treated as a thin film covering the agglomerate. Additionally, the spherical agglomerate is covered by a thin film of phosphoric acid water mixture. It is assumed that water is produced at the reaction site and then diffuses to the surface of the agglomerate, where a thin film of phosphoric acid and water mixture forms before leaving the cell through the porous media towards the gas channels.

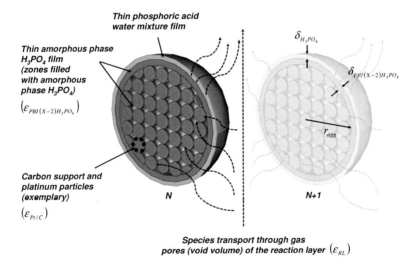

Fig.6. Schematic representation of the spherical agglomerate for reaction layer modeling.

The simplified Butler-Volmer equation describes the current density within both reaction layers. This equation is modified to account for the transport resistance of the species inside the spherical agglomerate, the amorphous phase phosphoric film, and the phosphoric acid water mixture film.

$$j_a = -a_a \cdot i^0_{PBI/(X-2)H_3PO_4,a} \cdot \left(\frac{c^g_{H_2}}{c^g_{H_2,ref}}\right)^{\gamma_a} \cdot e^{\alpha_a \cdot \frac{\cdot F}{R \cdot T_f} \cdot \eta_a}$$

$$j_c = (1-\varepsilon_{RL}) \cdot \left(1 - \frac{\varepsilon_{PBI/(X-2)H_3PO_4}}{\varepsilon_{agg}}\right) \cdot a_c \cdot i^0_{PBI/(X-2)H_3PO_4,c} \cdot \left(\frac{c^g_{O_2}}{c^g_{O_2,ref}}\right)^{\gamma_c} \cdot e^{-\alpha_c \cdot \frac{\cdot F}{R \cdot T_f} \cdot \eta_c} \cdot \frac{R \cdot T_f}{H_{O_2,PBI/(X-2)H_3PO_4}}$$

$$\times \xi_1 \cdot \frac{1}{1 + \xi_2 + \xi_3}$$

(45)

In [111] it was demonstrated that the logarithmic exchange current density of the oxygen reduction at platinum interfaced PBI/H_3PO_4 increases linearly with the amorphous phase phosphoric acid volume fraction. In this work, Eq.(46) is used to calculate the cathode side exchange current density.

$$i^0_{PBI/(X-2)H_3PO_4,c} = i^0_{H_3PO_4,c} \cdot \frac{1}{10^{4.16 \cdot (1-\varepsilon_{PBI/(X-2)H_3PO_4})}}$$

(46)

The local overpotential is defined as the difference between the solid-, and membrane-phase potential. This potential difference drives the current, keeping the electrochemical half-cell reactions continuous (Eq.(47)).

$$\eta_a = \phi_s - \phi_m - 0$$
$$\eta_c = \phi_s - \phi_m - E_{OCV}$$

(47)

The maximum equilibrium potential with respect to the temperature and partial pressures is calculated using the Nernst equation.

The effective surface area of the reaction layer is a function of the catalyst surface area per unit mass of the catalyst particle, the platinum loading, and the thickness of the reaction layer (Eq.(48)).

$$a_i = a_{Pt,i} \cdot \frac{m_{Pt,i}}{l_{RL,i}}$$

(48)

Gallart [112] summarized the available catalyst surface area per unit mass of the catalyst particle data and coupled this term to the platinum-to-carbon ratio, as shown in Eq.(49).

$$a_{Pt,i} = 7.401 \cdot 10^5 \cdot f_{Pt/C,i}^4 - 1.811 \cdot 10^6 \cdot f_{Pt/C,i}^3 + 1.545 \cdot 10^6 \cdot f_{Pt/C,i}^2$$

$$- 6.453 \cdot 10^5 \cdot f_{Pt/C,i} + 2.054 \cdot 10^5 \tag{49}$$

The analytical expression for the correction or effectiveness factor for the transport resistance of the species inside the spherical agglomerate is given by Eq.(50). This factor determines the extent to which the reactions are limited by species diffusion inside the agglomerate.

$$\xi_1 = \frac{1}{3 \cdot \phi_L^2} \cdot \left(3 \cdot \phi_L \cdot \coth(3 \cdot \phi_L) - 1 \right) \tag{50}$$

As shown by Eq.(51), the effectiveness factor depends on the Thiele modulus of the particular system.

$$\phi_L = \Gamma \cdot \sqrt{\frac{k_c}{D_{O_2,PBI/(X-2)H_3PO_4}^{agg}}} \tag{51}$$

The effectiveness factor and Thiele modulus in a HTPEM fuel cell model were discussed in [85]. For a low Thiele modulus, the effectiveness factor is approximately 1 and the reactant concentration is nearly constant within the pore (transport resistance is negligible). For a large Thiele modulus, the effectiveness factor can be approximated to the asymptote $1/\phi_L$. Under these conditions, the reactant concentration quickly falls to zero, and the reaction occurs primarily on the surface of the agglomerate [85]. The characteristic length scale of the spherical agglomerate (volume per surface area) is given by Eq.(52).

$$\Gamma = \frac{r^{agg}}{3} \tag{52}$$

The reaction rate is shown in Eq.(53).

$$k_c = \frac{\left(1 - \frac{\varepsilon_{PBI/(X-2)H_3PO_4}}{\varepsilon_{agg}} \right) \cdot a_c \cdot i_{PBI/(X-2)H_3PO_4,c}^0}{4 \cdot F \cdot c_{O_2,ref}^g} \cdot e^{-\alpha_c \frac{F}{R \cdot T_f} \eta_c} \tag{53}$$

The correction factor due to the amorphous phase phosphoric acid film covering the spherical agglomerate is given by Eq.(54).

$$\xi_2 = \frac{\delta^{PBI/(X-2)H_3PO_4}}{a^{PBI/(X-2)H_3PO_4} \cdot D_{O_2,PBI/(X-2)H_3PO_4}} \cdot \xi_1 \cdot k_c \tag{54}$$

The third correction factor in Eq.(45) accounts for the species transfer resistance due to the phosphoric acid water mixture film.

$$\xi_3 = \frac{\delta^{H_3PO_4}}{a^{H_3PO_4} \cdot D_{O_2,H_3PO_4}} \cdot \frac{H_{O_2,H_3PO_4}}{H_{O_2,PBI/(X-2)H_3PO_4}} \cdot \xi_1 \cdot k_c \tag{55}$$

The reference concentration of the species is calculated using Henry's law, incorporating the reference pressure and Henry's constant.

$$c^g_{O_2,ref} = \frac{p_0 \cdot x_{O_2,in}}{H_{O_2,PBI/(X-2)H_3PO_4}} \tag{56}$$

The effective diffusion coefficients within the amorphous phase phosphoric acid inside the agglomerate are calculated using the diffusion coefficient in the amorphous phase phosphoric acid film and the Bruggeman factor.

$$D^{agg}_{O_2,PBI/(X-2)H_3PO_4} = D_{O_2,PBI/(X-2)H_3PO_4} \cdot \left(\frac{\varepsilon_{PBI/(X-2)H_3PO_4}}{\varepsilon_{agg}}\right)^{1.5} \tag{57}$$

The diffusion coefficient within the amorphous phase phosphoric acid film is related to the diffusion coefficient for phosphoric acid. From a macroscopic point of view it was found that the oxygen diffusivity increases with increasing doping level. For highly doped PBI/H₃PO₄ systems, the diffusivity does not significantly vary from the data available for phosphoric acid [111]. In the present work, a similar expression is utilized, as presented in [77]. The Bruggeman relation, with an exponent of 1.8, is used to account for the agglomerate structure (Eq.(58)) [111].

$$D_{O_2,PBI/(X-2)H_3PO_4} = D_{O_2,H_3PO_4} \cdot \left(\varepsilon_{PBI/(X-2)H_3PO_4}\right)^{1.8} \tag{58}$$

The oxygen solubility in a PBI/H₃PO₄ system is related to the solubility in phosphoric acid [77]. The solubility is approximately four times higher than the values for phosphoric acid under the given conditions [113]. In [111], the authors stated that these higher values must be related to the presence of

PBI. Cheddie et al. [77] related the values of phosphoric acid to the values for different PBI/H$_3$PO$_4$ membrane doping levels using Eq.(59).

$$H_{O_2,PBI/(X-2)H_3PO_4} = \varepsilon_{PBI/(X-2)H_3PO_4}^{1.945} \cdot \left(\left(H_{O_2,H_3PO_4} \right) + 5.79 \cdot \left(1 - \varepsilon_{PBI/(X-2)H_3PO_4}^{1.8} \right) \right) \tag{59}$$

4.2.6 Kinetic parameters and transport properties in phosphoric acid

Microscopically, a PBI/H$_3$PO$_4$ system represents the combination of a crystalline region, the amorphous phase, and the free acid region. The excess phosphoric acid collects in the amorphous phase and acts as a concentrated phosphoric acid solution [111]. Because the water vapor equilibrates with the amorphous phase phosphoric acid within PBI/H$_3$PO$_4$, there should be only a minimal difference in the amorphous phase phosphoric acid concentration under similar conditions, regardless of the doping level. It follows that the oxygen reduction and hydrogen oxidation for a PBI/H$_3$PO$_4$ system are similar to those of phosphoric acid. Reported values of the exchange current density vary in the literature and depend on the temperature, phosphoric acid concentration, and dissolved oxygen concentration (Table 4).

Table 4

Published values for the exchange current density (values not corrected for temperature and pressure and measured using different catalysts, electrodes and different electrolytes).

Exchange current	H$_3$PO$_4$	Temperature	Reference
A cm^{-2}	wt.%	°C	
0.86-9.33·10^{-8}	100	150	[114]
3.8·10^{-9}	98	150	[115]
2.6·10^{-8}	98	150	[115]
1.0·10^{-8}	85	120	[116]
2.4·10^{-8}	85	136	[116]
1.6·10^{-8}	85	136.1	[117]
6·10^{-7}	85	150	[117]
8.0·10^{-8}	85	60	[118]
1.7·10^{-6}	85	150	[118]
2·10^{-6}	85	160	[118]
3.5·10^{-5}	85	150	[119]
3.8·10^{-9}	98	100	[120]
3.9·10^{-8}	98	125	[120]

$2.63 \cdot 10^{-8}$	98	150	[120]

Reported values for a PBI/H_3PO_4 membrane

$1.8 \cdot 10^{-9}$	$X = 4.5$	[111]
$2.9 \cdot 10^{-9}$	$X = 6$	[111]
$1.2 \cdot 10^{-8}$	$X = 8$	[111]
$2.4 \cdot 10^{-8}$	$X = 10$	[111]

In this work, the cathode side exchange current density for concentrated phosphoric acid is calculated using Eq.(60).

$$i^0_{H_3PO_4,c} = 1 \cdot 10^4 \cdot 10^{\left(-0.491 - 2193\frac{1}{T_f}\right)}$$ (60)

The transfer coefficients α_a and α_c depend on the temperature and phosphoric acid concentration. Table 5 lists the values reported in the literature.

Table 5

Published values for the transfer coefficient and Tafel slope.

Transfer coefficient	Slope	H_3PO_4	Temperature	Reference
	mV dec^{-1}	wt.%	°C	
	90-110	99	177	[121]
	90	100	190	[122]
	110-134	12.5-85	21	[123]
	120-200	7-95	25	[124]
0.47	125	85	25	[118]
0.61	120	85	100	[118]
0.67	125	85	150	[118]
0.51-0.66	115	85	25-136	[116]
0.94	90	96	160	[125]
0.53-0.68		85	25-70	[119]
0.60-1.01			100-175	[98]
Reported values for a PBI/H_3PO_4 membrane (different doping levels X)				
0.42-0.50	79-106	$X = 32$	130-180	[41]
0.81	104	$X = 10$	150	[111,131]
0.58-0.90		$X = 16$	100-175	[98]
0.91	92	$X = 4.5$	150	[98]

| 90-95 | $X = 32$ | 160 | [43] |
| 100-110 | $X = 32$ | 160-180 | [126] |

In this work, a cathode side transfer coefficient of 0.89 was found to produce the best results. At the anode side, a value of 0.5 is assumed, see, e.g., [85]. The oxygen solubility in the electrolyte and the oxygen diffusivity depend on the temperature and phosphoric acid concentration. Table 6 lists the values reported in the literature.

Table 6

Published values for the oxygen solubility and diffusivity.

Solubility	Diffusivity	H_3PO_4	Temperature	Reference
various units mol cm^{-3} (atm^{-1})	cm^2 s^{-1}	wt.%	°C	
$1-4 \cdot 10^{-7}$	$2.0-30.0 \cdot 10^{-6}$	85-96	100-150	[127]
$0.49-1.07 \cdot 10^{-7}$	$1.18-29.9 \cdot 10^{-6}$	98	25-150	[115]
$1.25-0.17 \cdot 10^{-7}$	$11.9-1.97 \cdot 10^{-6}$	7.4-85.6	23	[128]
$8.34-3.28 \cdot 10^{-7}$	$1.68-27.2 \cdot 10^{-6}$	100	75-150	[47]
$1.18-0.13 \cdot 10^{-6}$	$2.42-0.10 \cdot 10^{-6}$	5-100	25	[129]
$0.5 \cdot 10^{-6}$		95	150	[130]
	$2.2-4.2 \cdot 10^{-6}$	85	60-83	[130]
	$7.6 \cdot 10^{-7}$	85	25	[119]
Reported values for a PBI/H_3PO_4 membrane (different doping levels X)				
$0.57-1.13 \cdot 10^{-6}$	$2.8-8.0 \cdot 10^{-6}$	$X = 4.5-10$	150	[111,131]

Klinedinst et al. [127] noted that the oxygen diffusivity and solubility in phosphoric acid exhibit exponential reciprocal temperature dependencies over small temperature ranges. The activation energy for oxygen diffusion and the oxygen enthalpy of the solution vary with the phosphoric acid concentration. In this work, the transport properties of oxygen in phosphoric acid are related to the temperature and concentration, as presented in [77]. Similar empirical equations were used in [85,98].

$$D_{O_2,H_3PO_4} = 1 \cdot 10^{-9} \cdot e^{\left(\left(-192.55 \cdot \omega_{H_3PO_4}^2 + 323.55 \cdot \omega_{H_3PO_4} - 125.61 \right) + \frac{62010 \cdot \omega_{H_3PO_4}^2 - 105503 \cdot \omega_{H_3PO_4} + 40929}{T_f} \right)} \tag{61}$$

$$H_{O_2,H_3PO_4} = 1 \cdot 10^{-1} \cdot e^{\left(\left(257.13 \cdot \omega_{H_3PO_4}^2 - 431.08 \cdot \omega_{H_3PO_4} + 178.45 \right) + \frac{-93500 \cdot \omega_{H_3PO_4}^2 + 156646 \cdot \omega_{H_3PO_4} - 64288}{T_f} \right)} \tag{62}$$

The high values of the oxygen diffusion activation energy were assumed to be caused by the extensive hydrogen bond network and high viscosity of phosphoric acid [98]. It has been reported that the oxygen enthalpy of the solution decreases with increasing phosphoric acid concentration until it reaches negative values at 96wt.%. Thus, a smaller decrease in the solubility of oxygen in H_3PO_4 with temperature will occur as the acid concentration increases from 85wt.% to 95wt.%. Beyond this concentration, a slow increase in solubility with temperature occurs. Mamlouk [98] noted that the phosphoric acid concentration will vary greatly with the amount of water produced by the fuel cell due to ongoing electrochemical reactions (logarithmic relation). In [132], a correlating equation for phosphoric acid and water vapor pressure is given. Using experimental data, Sousa et al. [85] generated an equation that coupled the phosphoric acid concentration with the water vapor partial pressure (Eq.(63)).

$$x_{H_3PO_4} = \frac{\ln\left(x_{H_2O} \cdot p\right) + \dfrac{2765.1}{T_f} - 22.002}{\dfrac{-4121.9}{T_f} + 2.5929} \tag{63}$$

Choudhury et al. [106] used a similar equation to correlate the phosphoric acid concentration and water vapor pressure at a given temperature. The mole fraction of phosphoric acid is converted into a mass fraction with Eq.(64).

$$\omega_{H_3PO_4} = \frac{136 \cdot x_{H_3PO_4}}{111 \cdot x_{H_3PO_4} + 25} \tag{64}$$

Data for the hydrogen oxidation reaction at the anode side are not readily available. For the anode half-cell reaction, the exchange current density is taken to be $1 \cdot 10^8$ times the cathode side exchange current density, a value that is consistent with other published works, see, e.g., [77]. Within the different volume fractions of the agglomerate, the hydrogen diffusivity is taken to be two times the oxygen diffusivity and the hydrogen solubility is taken to be approximately four times the oxygen solubility for the same pressure, temperature, and phosphoric acid concentration. The behaviour is assumed to be similar to that of a water system [77].

$$D_{H_2,i} = 2 \cdot D_{O_2,i} \tag{65}$$

$$H_{H_2,i} = 4.44 \cdot H_{O_2,i} \tag{66}$$

4.2.7 PBI/H₃PO₄ sol-gel membrane modeling

The following equations describe the behaviour of a PBI/H₃PO₄ system with approximately 30-35 mol of phosphoric acid per PBI repeat unit. Two phosphoric acid molecules are bonded to PBI, whereas X-2 molecules remain free and tend to form the amorphous phase within the PBI/H₃PO₄ sol-gel membrane [133]. The amorphous phase phosphoric acid volume fraction within the membrane is calculated using Eq.(67).

$$\varepsilon_{PBI/(X-2)H_3PO_4,MEM} = \left(\frac{\frac{M_{PBI}}{M_{H_3PO_4}} + X}{X-2} \right)^{-1} \tag{67}$$

The PBI volume fraction within the membrane is calculated with Eq.(68).

$$\varepsilon_{PBI,MEM} = 1 - \varepsilon_{PBI/(X-2)H_3PO_4,MEM} \tag{68}$$

The amorphous phase phosphoric acid volume fraction contributes to the high membrane conductivity via a Grotthuss proton switching mechanism. The strong temperature dependency of the membrane conductivity is described using an Arrhenius approach [133].

$$\sigma = \frac{\sigma_0}{T_s} \cdot e^{\left(\frac{\Delta E^a(k_i, X)}{R \cdot T_s} \right)} \tag{69}$$

The pre-exponential term is assumed to be independent of the operating temperature and decreases with increasing doping levels. The concentration of the mobile species in Eq.(70) should change with the doping level [133].

$$\sigma_0 = \left(\frac{z^2 \cdot F^2}{R} \right) \cdot \alpha_h \cdot \upsilon_0 \cdot d^2 \cdot c \cdot e^{\frac{\Delta S + \Delta S^f}{R}} \tag{70}$$

The activation energy of the conductivity depends on multiple factors, including the membrane doping level and polymer backbone structure. Relatively few datasets are available in the literature for PBI/H₃PO₄ sol-gel membranes [126,134]. In this work, the values for the pre-exponential factor and activation energy are taken from [41].

4.2.8 Fluid-(gas)-, and solid-phase properties and material correlations

A woven-type gas diffusion layer (similar to E-tek – ELAT® products [135]) is considered herein. The layer consists of a void volume fraction and solid-phase volume fraction (Eq.(71)).

$$\varepsilon_{GDL}^{0} + \varepsilon_{GDL}^{s} = 1 \tag{71}$$

The structure of the reaction layer is more complex. Different volume fractions are needed to calculate the effective properties. The volume fraction of platinum and carbon is calculated from Eq.(72), considering the thickness and catalyst properties [136].

$$\varepsilon_{Pt/C} = \frac{m_{Pt}}{l_{RL} \cdot f_{Pt/C}} \cdot \left(\frac{f_{Pt/C}}{\rho_{Pt}} + \frac{1 - f_{Pt/C}}{\rho_{C}} \right) \tag{72}$$

The fraction of the volume occupied by the amorphous phase phosphoric acid within the agglomerate is calculated using Eq.(73) [136].

$$\varepsilon_{PBI/(X-2)H_3PO_4} = \frac{m_{Pt}}{l_{RL} \cdot f_{Pt/C}} \cdot \left(\frac{f_{PBI/(X-2)H_3PO_4}}{\left(1 - f_{PBI/(X-2)H_3PO_4}\right) \cdot \rho_{PBI/(X-2)H_3PO_4}} \right) \tag{73}$$

The volume fraction of the agglomerate can be calculated with the following equation.

$$\varepsilon_{agg} = \varepsilon_{Pt/C} + \varepsilon_{PBI/(X-2)H_3PO_4} \tag{74}$$

The porosity of the reaction layer is calculated with Eq.(75).

$$\varepsilon_{RL} = 1 - \varepsilon_{agg} \tag{75}$$

The surface area of the agglomerate per unit volume of reaction layer is calculated as follows.

$$a^{PBI/(X-2)H_3PO_4} = 4 \cdot \pi \cdot n^{agg} \cdot \left(r^{agg} + \delta^{PBI/(X-2)H_3PO_4} \right)^2 \tag{76}$$

The surface area of the agglomerate per unit volume including the phosphoric acid water film mixture per unit volume of the reaction layer is calculated from Eq.(77).

$$a^{H_3PO_4} = 4 \cdot \pi \cdot n^{agg} \cdot \left(r^{agg} + \delta^{PBI/(X-2)H_3PO_4} + \delta^{H_3PO_4} \right)^2 \tag{77}$$

The number of agglomerates per unit volume is given by Eq.(78).

$$n^{agg} = \frac{3 \cdot \varepsilon_{agg}}{4 \cdot \pi \cdot \left(r^{agg} + \delta^{PBI/(X-2)H_3PO_4} \right)^3} \tag{78}$$

The amorphous phase phosphoric acid film thickness is defined as

$$\delta^{PBI/(X-2)H_3PO_4} = \sqrt[3]{\left(r^{agg} \right)^3 \cdot \left(1 + \frac{\varepsilon_{PBI/(X-2)H_3PO_4}}{\varepsilon_{Pt/C}} \right)} - r^{agg} \tag{79}$$

The phosphoric acid film thickness is calculated using Eq.(80).

$$\delta^{H_3PO_4} = \sqrt[3]{\left(r^{agg} + \delta^{PBI/(X-2)H_3PO_4} \right)^3 \cdot \left(1 + \frac{\varepsilon_{H_3PO_4}}{\varepsilon_{PBI/(X-2)H_3PO_4}} \right)} - \left(r^{agg} + \delta^{PBI/(X-2)H_3PO_4} \right) \tag{80}$$

The binary diffusion coefficients are calculated for all pairs of species in the fluid-(gas)-phase mixture using Eq.(81) and are known to vary with pressure and temperature.

$$\tilde{D}_{ij} = k \cdot \frac{T_f^{1.75}}{P \cdot \left(v_i^{\frac{1}{3}} + v_j^{\frac{1}{3}} \right)^2} \cdot \sqrt{\frac{1}{M_i} + \frac{1}{M_j}} \tag{81}$$

The effective porous media binary diffusivities are calculated using the appropriate void volume fractions and the Bruggeman relationship with an exponent of 1.5, which is consistent with other works.

$$\tilde{D}_{ij}^{eff} = \tilde{D}_{ij} \cdot \left(\varepsilon_{GDL,RL}^0 \right)^{1.5} \tag{82}$$

The density of the fluid-(gas)-phase mixture within the gas channel and the porous media are calculated using the mole fraction and molar mass of the gas species.

$$\rho_i = \sum_i x_i \cdot M_i \cdot \frac{P}{R \cdot T_f} \tag{83}$$

The fluid-(gas)-phase properties are taken from the material library provided by the used software and depend on the fluid-(gas)-phase temperature.

The thermal conductivity of the membrane is calculated using the thermal conductivities of PBI and amorphous phase phosphoric acid.

$$k_{MEM} = k_{PBI} \cdot \left(1 - \varepsilon_{PBI/(X-2)H_3PO_4,MEM}\right) + k_{PBI/(X-2)H_3PO_4} \cdot \varepsilon_{PBI/(X-2)H_3PO_4,MEM} \tag{84}$$

Based on the data provided by Turnbull [137], the thermal conductivity data is extrapolated for a higher temperature and phosphoric acid concentration with Eq.(85).

$$k_{PBI/(X-2)H_3PO_4} \approx k_{H_3PO_4} \tag{85}$$

$$k_{H_3PO_4} = \left(\frac{11.727 + 0.01864 \cdot T_s - 0.02169 \cdot c_{H_3PO_4} - 0.0000338 \cdot c_{H_3PO_4} \cdot T_s}{1 \cdot 10^4}\right) \cdot 418.68$$

The remaining gas mixture properties are calculated using average-based mole or mass fractions. The properties of the individual species are directly taken from the material database of the software.

4.3 Boundary conditions

The boundary conditions are defined according to the experimental investigation. The inlet flow should be fully developed as it enters the cell. An additional pressure variable (weak contribution / additional degrees of freedom (DOF)) is used to define a laminar inflow condition (volume per unit time).

$$L_{entr} \cdot \nabla_t \cdot \left(p \cdot I - \eta \cdot \left(\nabla_t \cdot u + (\nabla_t \cdot u)^T\right)\right) = -n \cdot p_{entr}$$
$$\nabla_t \cdot u = 0 \tag{86}$$

At the outlet, a pressure condition is used (no viscous stress).

$$\eta\left(\nabla u + (\nabla u)^T\right) \cdot n = 0$$
$$p = p_0 \tag{87}$$

At the bipolar-plate walls (gas channel walls), a no-slip boundary condition is defined for momentum transport

$$u = 0 \tag{88}$$

At all other boundaries, a continuity boundary condition is set.

For the mass (species) balance, the mass fractions are defined according to the given gas compositions at the inlet of the gas channels.

$$\omega_i = \omega_{i,0} \tag{89}$$

At the outlet of the gas channels, a convective flux is set according to Eq.(90).

$$n \cdot \left(-\rho \cdot \omega_i \cdot \sum_k \tilde{D}_{ij}^{eff} \cdot \left(\nabla x_k + (x_k - \omega_k) \cdot \frac{\nabla p}{p}\right) + D_i^T \cdot \frac{\nabla T_f}{T_f} + \rho \cdot \omega_i \cdot u_i\right) = 0 \tag{90}$$

At all other boundaries, a continuity or insulation boundary condition is used.

For the charge balance, the anode solid-phase potential is set to 0 V. The cell operating voltage is defined at the cathode side gold-plated copper current collector.

$$\phi_{s,a} = 0$$
$$\phi_{s,c} = U_{cell} \tag{91}$$

For the remaining boundaries, either an electric insulation or continuity boundary condition is used.

The following boundary conditions are applied for the fluid-(gas)-, and the solid-phase temperatures. The set solid-phase temperature is defined at the anode, and cathode side gold-plated copper current collector boundaries.

$$T_s = T_{s,0} \tag{92}$$

A heat transfer coefficient is used to account for the heat absorbed by the gases in both energy transport equations.

$$-n \cdot \left(-k\nabla T_{s,f}\right) = q_0 + h_{boundary} \cdot \left(T_{boundary} - T_{s,f}\right) \tag{93}$$

For the remaining boundaries, an insulation and symmetry or continuity boundary condition is used.

The fluid-(gas)-phase temperature is defined at the inlet of the gas channels, and a convective flux boundary condition is utilized at the gas channel outlets respectively.

$$T_f = T_{f,0} \tag{94}$$

$$-n \cdot \left(-k\nabla T_f\right) = 0 \tag{95}$$

Similar to the previous settings, a symmetry, continuity, or insulation boundary condition is used at all other boundaries.

4.4 Initial conditions

In addition to the modeling and simulation performed with CFD software, simple analytical calculations are used to estimate the order of magnitude of the quantities. These simple calculations aid in interpreting the results in terms of consistency and provide the initial conditions for the application modes (Table 7).

Table 7

Used initial conditions for the application modes.

Application mode	Parameter and value	Unit	Note
Incompressible	$U = 0$	m s^{-1}	used at anode and cathode side
Navier-Stokes	$p = 1.01325 \cdot 10^5$	Pa	used at anode and cathode side
Conductive media DC	$\phi_s = 0$	V	used at anode side
	$\phi_s = U_{cell}$	V	used at cathode side
	$\phi_m = 0$	V	used at anode and cathode side
Maxwell-Stefan	$\omega_{H_2}, \omega_{H_2Og,a}$		used at anode side
diffusion and convection	$\omega_{O_2}, \omega_{H_2Og,c}, \omega_{N_2}$		used at cathode side
General heat transfer	$T_s = T_{s,0}$	K	used at anode and cathode side
	$J = 0$	W m^{-2}	used at anode and cathode side
	$T_f = T_{f,0}$	K	used at anode and cathode side
	$J = 0$	W m^{-2}	used at anode and cathode side

4.5 Assumptions and simplifications

The following assumptions are generally adopted when modeling a HTPEM fuel cell (e.g., assumptions 2, 5, 9, and 12). Some of the simplifications are directly related to steady-state operating conditions (assumption 1) and the modeling level, the purpose, and the outcome of the study (e.g., contact resistances are mostly neglected in such studies but must be included when modeling cell compression or structural mechanics). Other simplifications are based on material properties and material casting methods (e.g., assumptions 6, and 7). The following assumptions and simplifications are used in this work:

1) Steady-state operating conditions.
2) Continuity is prescribed at the internal boundaries (interfaces); all contact resistances are neglected.
3) Some material properties are assumed to be isotropic and macrohomogeneous, whereas others account for different material properties along the x-, y-, and z-axis.
4) A microporous layer and its influence on the quantities distributions are not explicitly considered in this model.
5) All product water is assumed to be vaporous (no phase change) and to leave the cell in vapor form.
6) Gas and water crossover through (or water uptake by) the PBI/H$_3$PO$_4$ sol-gel membrane is neglected.

7) The initial concentration of the amorphous phase phosphoric acid volume fraction inside the reaction layer and membrane is assumed to be 85wt.% at room temperature and pressure.

8) Only orthophosphoric acid is considered herein, other forms of phosphoric acid (e.g., pyrophosphoric acid) are neglected.

9) The membrane is considered to be a system of PBI, phosphoric acid and water.

10) Low velocities (low Reynolds numbers) are expected in the gas streams. Laminar inflow conditions are assumed.

11) All agglomerates are assumed to be geometrically identical, spherical in shape, and of the same radii.

12) There is no heat transfer towards the surroundings.

13) Ideal gas mixtures are used.

4.6 Modeling parameters

Table 8 summarizes the modeling parameters and constants used in this work. Some of these values were directly provided by the manufacturers, textbooks, and publications. Material properties and thermodynamic data are taken from the software database. The reference operating conditions are described in Table 12.

Table 8

Modeling parameters and constants.

Parameter	Value	Unit	Reference	Note
F	96,485	A s mol^{-1}		
h_{T_s/T_f}	$5 \cdot 10^5$	W m^{-3} K^{-1}	[102]	used at anode and cathode
K	$3.16 \cdot 10^{-8}$			
k_{CU}	400	W m^{-1} K^{-1}	[71]	
k_{BPP}	20	W m^{-1} K^{-1}	[138]	
$k_{GDL,x-y}$	2.2	W m^{-1} K^{-1}		assumed to be ten times $k_{GDL,z}$
$k_{GDL,z}$	0.22	W m^{-1} K^{-1}	[143]	
$k_{p,GDL}$	$1 \cdot 10^{-12}$	m^2	[139]	assumption based on the literature cited
$k_{p,GDL}$	$1 \cdot 10^{-13}$	m^2	[139]	assumption based on the literature cited
k_{RL}	0.5	W m^{-1} K^{-1}		assumed
M_{H_2}	0.002	kg mol^{-1}		
M_{O_2}	0.032	kg mol^{-1}		

M_{N_2}	0.028	kg mol^{-1}		
M_{H_2O}	0.018	kg mol^{-1}		
$M_{H_3PO_4}$	0.098	kg mol^{-1}		
M_{PBI}	0.308	kg mol^{-1}		
r_{agg}	25	µm	[112]	
R	8.314	J mol^{-1} K^{-1}		
$T_{f,0}$	21	°C		
$T_{s,0}$	160	°C		
X	32			assumption based on [41]
α_a	0.5		[85]	
α_c	0.89			assumption based on [41]
$\varepsilon_{H_3PO_4}$	0.18			assumption
$\ln(\sigma_0 \cdot T_s)$	9.5211	S K cm^{-1}	[41]	
σ_0	13,466	S K m^{-1}	[41]	
σ_{CU}	10,000	S m^{-1}	[71]	
$\sigma_{BPP,x-y}$	20,000	S m^{-1}	[138]	
$\sigma_{BPP,z}$	4,170	S m^{-1}	[138]	
$\sigma_{GDL,x-y}$	2,100	S m^{-1}		assumed to be ten times $\sigma_{GDL,z}$
$\sigma_{GDL,z}$	210	S m^{-1}	[140]	assumption based on the literature cited
$\sigma_{m,RL}$	13	S m^{-1}		assumed
$\sigma_{s,RL}$	400	S m^{-1}	[141]	
ν_{H_2}	$7.07 \cdot 10^{-6}$		[142]	
ν_{O_2}	$16.6 \cdot 10^{-6}$		[142]	
ν_{N_2}	$17.9 \cdot 10^{-6}$		[142]	
ν_{H_2O}	$12.7 \cdot 10^{-6}$		[142]	
ρ_C	2,000	kg m^{-3}	[71]	
$\rho_{PBI/(X-2)H_3PO_4}$	1,698	kg m^{-3}	[111]	
ρ_{Pt}	21,500	kg m^{-3}	[71]	
ΔE^a	18,484	J mol^{-1}	[41]	

5. Solving the model

The commercially available finite element software COMSOL Multiphysics® v3.5a (64 bit version), along with the CAD import module, chemical engineering module, and heat transfer module, was used for all modeling and simulation. Predefined application modes were used to solve the governing equations. Table 9 highlights the application modes and dependent variables used in this work. The operating system was Windows XP (64 bit version). The software was installed on a HP XW8600 workstation equipped with two INTEL Xeon 5460 quadcores and 64 GB DDR2 RAM (16 x 4 GB RAM).

Table 9

Application modes used for modeling and simulation.

Application mode name	Dependent variables	Application modes properties
Incompressible Navier-Stokes (2x)	u, v, w, p, p_{inl}	Element type: Lagrange – $P_1 P_2$
		Corner smoothing: off
		Weak constraints: off
		Constraint type: ideal
Conductive media DC (2x)	ϕ_s, ϕ_m	Element type: Lagrange – $P_1 P_2$
		Weak constraints off
		Constraint type: ideal
Maxwell-Stefan diffusion and convection (2x)	ω_{O_2}, ω_{H_2Ogc}, ω_{N_2} ω_{H_2}, ω_{H_2Oga}	Element type: Lagrange – quadratic
		Weak constraints off
		Constraint type: ideal
General heat transfer (2x)	T_s, J_s T_f, J_f	Element type: Lagrange – quadratic
		Weak constraints off
		Constraint type: ideal

5.1 Meshing

The three-dimensional model geometry was created using CAD software and imported as a neutral standard (*.iges) into COMSOL Multiphysics®. The import tolerance was set to $1 \cdot 10^{-5}$ m, and the imported data were repaired and verified if necessary. All subdomains of the model geometry were separately discretized using tetrahedral and prism elements in mesh case 1 (multigrid level 1). Meshing needed to be performed carefully to minimize the number of degrees of freedom and the memory requirements and to improve the quality of the mesh.

Table 10

Meshing details of the different three-dimensional model geometries.

Geometry	Number of subdomain elements	Number of boundary elements	Additional information
Geometry with type I flow-field	Mesh case 1 697,463 (total) 299,819 (tetrahedral) 397,644 (prism)	Mesh case 1: 258,782 (total) 214,970 (triangular) 43,812 (quadrilateral)	Mesh case 1: Element quality: 0.014 Ratio: $2.29 \cdot 10^{-6}$
	Mesh case 2: 218,552 (tetrahedral)	Mesh case 2: 93,502 (triangular)	Mesh case 2: Element quality: 0.012 Ratio: $7.80 \cdot 10^{-5}$
	Mesh case 3: 109,246 (tetrahedral)	Mesh case 3: 48,136 (triangular)	Mesh case 3: Element quality: 0.005 Ratio: $1.41 \cdot 10^{-5}$
Geometry with type II flow-field	Mesh case 1: 774,3687 (total) 324,956 (tetrahedral) 449,412 (prism)	Mesh case 1: 280,391 (total) 236,549 (triangular) 43,842 (quadrilateral)	Mesh case 1: Element quality: 0.011 Ratio: $1.64 \cdot 10^{-6}$
	Mesh case 2: 189,652 (tetrahedral)	Mesh case 2: 82,511 (triangular)	Mesh case 2: Element quality: 0.019 Ratio: $1.00 \cdot 10^{-4}$
	Mesh case 3: 83,926 (tetrahedral)	Mesh case 3: 37,239 (triangular)	Mesh case 3: Element quality: 0.013 Ratio: $2.32 \cdot 10^{-4}$
Geometry with type III flow-field	Mesh case 1: 815,312 (total) 353,342 (tetrahedral) 461,970 (prism)	Mesh case 1: 288,540 (total) 242,097 (triangular) 46,443 (quadrilateral)	Mesh case 1: Element quality: 0.008 Ratio: $3.19 \cdot 10^{-6}$
	Mesh case 2: 247,209 (tetrahedral)	Mesh case 2: 102,613 (triangular)	Mesh case 2: Element quality: 0.017 Ratio: $2.34 \cdot 10^{-5}$
	Mesh case 3: 114,976 (tetrahedral)	Mesh case 3: 49,968 (triangular)	Mesh case 3: Element quality: $8.30 \cdot 10^{-4}$ Ratio: $7.48 \cdot 10^{-6}$

The manual mesh procedure is as follows. The top boundaries of the gas channel subdomains were meshed first, followed by a swept mesh operation. The same procedure was repeated for the bipolar-plate subdomains. For the porous media, the gas diffusion layer subdomains were meshed first, followed by reaction layer subdomains (swept mesh operation). The meshes of the porous media subdomains were refined to account for the expected local quantity gradients. The membrane subdomain was meshed with tetrahedral elements only because the two bipolar-plate subdomains (and flow-field subdomains) are turned 180° with respect to each other (see Fig.5). This set-up resulted in a non-conformal mesh at both membrane-reaction layer boundaries. The remaining subdomains (gold-plated copper current collector subdomains and the upper part of the bipolar-plate subdomains) were meshed exclusively with relatively coarse tetrahedral elements because only scalar variables were solved within these particular subdomains. After completing mesh case 1, two additional coarser mesh cases (multigrid levels 2 and 3) were manually generated to use a multigrid solver. Table 10 lists the meshing details of the three different model geometries for the mesh cases 1, 2, and 3. Based on the number of mesh elements, the number of dependent variables, and the application mode properties, the total number of degrees of freedom was calculated to be 12,665,886 for the geometry with type I flow-field, 14,077,677 for the geometry with type II flow-field, and 14,557,298 for the geometry with type III flow-field. Fig.7 provides an example of the three mesh cases (three multigrid levels) used in this work.

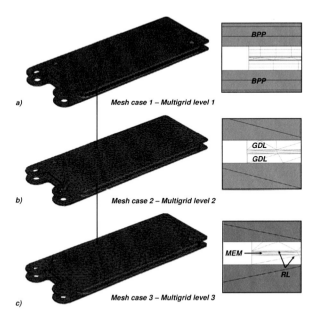

Fig.7. Manually generated mesh cases (3 multigrid levels shown in (a), (b), and (c)) (x-y-z-plane) and detailed view of the meshed membrane electrode assembly (y-z-plane).

5.2 Solution procedure and convergence behaviour

The different solvers were carefully tuned to reach a converged solution. Due to the high number of unknowns, a large amount of memory was needed, even when using iterative solvers. The best possible initial conditions were generated using analytical calculations and a series of parametric dummy simulations. The solution procedure was rather complex due to the strong multiphysics couplings of the transport equations. Initially, three solution steps were defined and the solved model was completely decoupled (e.g., assuming isothermal and isobaric operating conditions, and constant material properties).

Table 11

Used solution steps and solver settings.

Solution step	Solver settings	Note
Step 1: u, v, w p, p_{inl}	Parametric direct solver PARDISO	High gas viscosity and low volume flow rate at gas channel inlets used
Step 2: ϕ_s, ϕ_m $\omega_{O_2}, \omega_{H_2Og,c}, \omega_{N_2}$ $\omega_{H_2}, \omega_{H_2Og,a}$	Segregated group 1 (ϕ_s, ϕ_m): Iterative 3 level multigrid GMRES solver with SOR presmoother and SORU postsmoother and direct coarse solver (PARDISO) Segregated group 2 ($\omega_{O_2}, \omega_{H_2Og,c}$): Direct solver PARDISO Segregated group 3 (ω_{H_2}): Direct solver PARDISO	Low and constant overpotential used
Step 3: T_s, J_s T_f, J_f	Segregated group 1 (T_s, J_s): Iterative 3 level multigrid GMRES solver with SOR presmoother and SOR(U) postsmoother and direct coarse solver (PARDISO) Segregated group 2 (T_f, J_f): Direct solver PARDISO	
Step 4:	Step 1, 2, 3 repeated until all convergence criteria are satisfied	

The Navier-Stokes equations were first solved with a high gas viscosity value and a low volume flow rate. In the next step, the charge and mass (species) transport equations were solved using constant low, overpotentials. In the last step, the fluid-(gas)-, and the solid-phase temperature distributions were solved. The multiphysics couplings were increased in each solution step, and the properties were tuned back to their original values. The above steps were repeated in an iterative fashion while continuously monitoring the simulation results until the individual convergence criteria of $1 \cdot 10^{-6}$ was satisfied. Table 11 lists the solver settings used in this work. The UMFPACK direct solver was used in some cases instead of the PARDISO direct solver. For the settings given in Table 11, a converged solution was typically returned after less than 20 iterations. The number of iterations and the calculation time required to meet the convergence criteria were strongly dependent on the cell operating voltage. Overall, the time required to calculate a typical I-V-curve for 12 different load currents was approximately 48 h of clock time. The system matrix factorization generally took longer than the matrix solution process itself.

6. The segmented HTPEM fuel cell

Over the past few decades, various invasive and noninvasive fuel cell diagnostic methods have been developed. Optical measurements [144-146] and neutron radiography [147-151] have been used to visualize and quantify liquid water formation and water transport within different components of a working cell. Other methods include (soft) X-ray radiography and imaging [152-154] and nuclear magnetic resonance imaging [155]. Additionally, segmented temperature, segmented current density, and segmented EIS measurement techniques contribute to our understanding of the operational behaviour of fuel cells.

6.1 Segmented measurements – Literature review

6.1.1 Temperature measurements

In HTPEM fuel cells, several current issues are directly related to temperature cycling, tempering concepts, and heat-up and shut-down procedures [51-53,156,157]. Especially when a fuel cell is operating at a very high current density, a significant amount of research and material design (e.g., thermal conductivity of the gas diffusion layer material) is needed. In situ temperature measurements may be realized with embedded (micro)thermocouples, as demonstrated by Mench et al. [158], who placed several thermocouples between two sheets of Nafion®, and Wilkinson et al. [159]. He et al. [160] worked with thin film gold thermistors to measure the temperature evolution, whereas pyrometry was used in [161] and micro-electromechanical system sensors were employed in [162,163]. Other groups have used Bragg fiber temperature sensors (optical measurement methods) [164] and laser absorption spectroscopy [165] or have estimated the internal temperature by measuring the external temperature [166]. In [167], the authors implemented thermal sensors (calibrated with an uncertainty of ±0.6°C) based on the principles of the lifetime-decay method of phosphor thermometry to measure the temperature inside a LTPEM fuel cell. Optical sensors were used to measure the temperature at the interface of the gas diffusion layer and the cathode side bipolar-plate under different operating conditions and different load currents. Hakenjos et al. [168,169] used a transparent cell (a zinc selenide window) and measured the temperature with an infrared camera. Lebæk et al. [170] published in situ temperature measurements of a HTPEM fuel cell. In total, 16 type-T thermocouples were embedded within the land area of the bipolar-plates to investigate the temperature behaviour for various modes of operation. The temporal evolution of the temperature for different stoichiometric flow rates was highlighted. The sensors had an effect on the fuel cell performance, which was unexpected because the sensor positions were designed to not disturb the cell performance. Solid-phase temperature measurements of a HTPEM fuel cell with three types of flow-fields were presented in [171], and an influence of the fluid-(gas)-, on the solid-phase temperature distribution was confirmed.

6.1.2 Current density measurements

Reviews of the scientific literature on various current density measurement techniques have been provided by Schulze et al. [172], Sauer et al. [173], Hartnig et al. [174], and Pérez [175], among others. Studies on segmented measurements in HTPEM fuel cells are scarce and relatively new. Schaar [82] used a commercially available measurement system from the company S++ [176] to measure the current density distribution over a relatively large active area of 200 cm^2 and compared the experimental results to those obtained by simulations. In [177], a two channel parallel serpentine flow-field was divided into several segments. The current density was measured over an active area of 10 cm^2 for various operating conditions. A linear shift in current density was observed for selected segments therein as a consequence of the anode gas contaminants. In [178], a 25 cm^2 membrane electrode assembly was used to perform segmented current density measurements. The results showed that the current density distribution is primarily determined by the stoichiometric flow rates at the cathode side. The results also indicated that a reduction in the air flow leads to changes in the current density distribution. Lobato et al. [179] presented an interesting study in which the current density distributions for different types of flow-fields were compared with each other. Results were presented for oxygen and air as the cathode gas. A theoretical-practical study of a HTPEM fuel cell with an active area of 50 cm^2 was published in [180], presenting the influence of the flow-field geometry. Segmented current density measurements of a HTPEM fuel cell with a 50 cm^2 membrane electrode assembly were presented in [181,182]. The highest solid-phase temperature and current density were recorded in the region of the air inlet for most operating conditions.

6.1.3 EIS measurements

EIS is a well established technique for investigating LTPEM fuel cells [183,184]. Because HTPEM fuel cell technology is relatively new, only a few nonsegmented and segmented EIS datasets have been reported. Jalani et al. [126] used EIS measurements to obtain a detailed view of the PBI/H$_3$PO$_4$ sol-gel membrane processes under different operating conditions including variable operating temperatures, stoichiometric flow rates, and anode humidification. A high frequency intercept of 0.10 Ω cm^2 was reported. By varying the stoichiometric flow rates at the cathode side, the observed change in performance and impedance behaviour indicated that a variation in oxygen partial pressure from inlet to outlet had a profound effect on the global responses in voltage and impedance. The impedance signatures developed during fuel starvation exhibited a 45° line with a similar signature when the cell was electrically shorted. Zhang et al. [134] used EIS and cyclic voltammetry to obtain the exchange current density and the activation energy, respectively, for both half-cell reactions. Herein, the ohmic contribution was slightly below 0.10 Ω cm^2 when the cell operated at 160°C. The used equivalent circuit consisted of two resistor capacitor (constant phase element – CPE) pairs and one resistor to analyze the charge and gas transfer resistances. Additionally, different methods for improving the gas diffusion processes in HTPEM fuel cells were highlighted. Andreasen et al. [185] collected extensive EIS measurements to examine the poisoning effects on HTPEM fuel cell using reformate gas. The impedance was evaluated for different operating temperatures, currents, and anode gas compositions using equivalent circuit modeling. Their model consisted of two resistor CPE pairs (with the exponent fixed to 0.85), one resistor capacitor pair, one resistor inductance pair and one resistor to distinguish between the high, intermediate, and low frequency loops. The parameters for numerous operating conditions were discussed in great detail. A value of approximately 0.20 Ω cm^2 was reported for the high frequency resistance at 160°C. The same authors also published a study on characterizing HTPEM fuel cell stacks using EIS measurements [186]. Equivalent circuit modeling was performed, and the parameters and transfer functions were discussed. Another published work highlighted the effects of temperature on membrane hydration and catalyst particle agglomeration [187]. Mamlouk et al. [188] characterized the effect of electrode parameters, such as the platinum catalyst content in wt.%, acid doping in both PBI and PTFE based electrodes, and catalyst heat treatment, on the kinetic and mass transport characteristics. The influence of the load current, operating temperature and oxidant gas on the response was demonstrated and interpreted using different equivalent circuits. Bandlamudi [41] used EIS to analyze the impedance of HTPEM fuel cells under different compressions and presented Nyquist plots of highly and lowly doped membranes under different load currents. The evolution of various resistances was studied at different operating temperatures and load currents using a simplified Randles model. Hu et al. [189,190] performed EIS measurements over a 500 h continuous test and found that the degradation was primarily caused by platinum particle agglomeration. Schaltz et al. [191] and Jespersen et al. [192] reported EIS measurements and performed equivalent circuit modeling using two resistor CPE pairs and one resistor to distinguish between the charge and gas transfer resistance. A high frequency resistance of 0.20 Ω cm^2 was reported at 160°C. Kurz [99] performed EIS measurements on a HTPEM fuel cell stack under different operating conditions. The shape and size of the spectra were nearly identical for operating temperatures of 140°C and 170°C. The high frequency resistance was measured to be 0.102 Ω m^2 and 0.84 Ω m^2 under typical HTPEM fuel cell operating conditions. Segmented EIS measurements of LTPEM fuel cells were published by Andreaus et al. [193] and Brett at al. [194] who presented a method for measuring the spectroscopy response of the localized electrochemical impedance over a wide frequency range using a Solartron 1260 hardware, a multiplexer, a rack of 10 loads, and individual current lines. Hakenjos et al. [168] presented segmented current density and EIS measurement results for a LTPEM fuel cell. These measurements were performed using a multichannel frequency response analyzer with a multichannel potentiostat. In [195], a similar measurement set-up was used (Solartron 1254 frequency response analyzer with two 1251 multichannel extensions) and enabled the simultaneous measurements of the impedance spectra for single cells in a fuel cell stack. Schneider et al. [196-199] published several detailed works on segmented EIS measurements in LTPEM fuel cells, including investigation on the water balance under different operating conditions, direct measurements of the local current density at submillimeter resolution, local cyclic voltammetry, and segmented EIS measurements in the channel and land areas. Hogarth et al. [200] used a 20 cm^2 segmented fuel cell to investigate the performance of hydrogen-air using two different flow-fields. Each segment was connected to a separate current line and voltage sensor, and all measurements were taken after setting the local voltage of each segment constant to prevent any cross currents in the electrode. Gerteisen et al. [201] published a 50 channel characterization system for PEM fuel cells. This system was capable of both traditional electrochemical measurements and EIS measurements. The system relied on dedicated potentiostats (550-R from MaterialsMates) to control the current and voltage with independent frequency response analyzers (520-4 from MaterialsMates). In another work [202], the same authors used a segmented cell consisting of 49 segments to measure the local current density and high frequency resistance for various operating conditions using a three-channel serpentine flow-field. Based on the experimental results published by Schneider et al. [203,204], Kulikovsky [205] recently introduced a model for a segmented electrode for LTPEM fuel cells and discussed the high and low frequency loops in the impedance spectra. It was shown that a slight difference in the local impedance spectra could occur when exciting only one segment while all other segments were not excited. Reshetenko et al. [206-210] published studies on LTPEM fuel cells discussing the effects of back pressure, stoichiometric flow rates, gas humidification, media material porosity, gas composition, and flow-field design on the current density distribution. The authors used different equivalent circuit models to evaluate the parameters. The ability to measure the segmented current density distribution and EIS of HTPEM fuel cells is new, and only a few studies are currently available. Bergmann et al. [178] performed segmented EIS measurements on a HTPEM fuel cell (25 cm^2 MEA) using a 50 channel (synchronized) potentiostat-impedance spectrometer. Various operating conditions were analyzed, up to an operating temperature of 200°C. The results indicated that a reduction in the air flow leads to changes in the impedance spectra. The impedance spectra displayed more pronounced second loops for cell segments towards the air outlet. The high frequency resistance did not vary with the stoichiometric flow rates at the cathode side and did not vary between the investigated segments. Similar results were presented in [211] for a HTPEM fuel cell with straight flow-fields under various operating conditions. Several current density distributions were evaluated and the EIS spectra were fitted using equivalent circuits.

6.2 Requirements, manufacturing and assembling

For the segmented HTPEM fuel cell set-up, all of the materials were high temperature and chemically stable. It was also important to maintain the design as similar as possible to the original nonsegmented design. After the CAD work was complete, the different components of the segmented cathode side bipolar-plate were manufactured in the workshop. Fig.8 shows the different steps for this process. First, the insulation matrix was manufactured out of a polyetheretherketone (substrate plate), followed by bipolar-plate segments from Eisenhuth, Germany [138] and the gold-plated copper current collector segments. In the next step, the segmented cathode side bipolar-plate was assembled with custom made O-rings (FKM material) and all of the segments. In the last step, the flow-field was milled into the blank bipolar-plate material.

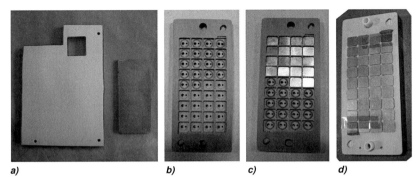

Fig.8. Segmented cathode side bipolar-plate during manufacturing and assembling. Polyetheretherketone and the bipolar-plate material (a). Insulation matrix with O-rings (b). Insulation matrix with O-rings and gold-plated copper current collector segments (c). Segmented cathode side bipolar-plate without the flow-field (d).

After the manufacturing process, the complete segmented cathode side bipolar-plate was disassembled and all components were cleaned. During the manufacturing and assembling processes, the surfaces of all of the components were scanned several times using a measuring system from Fries Research & Technology, Germany [212]. Different scans were performed to evaluate the topology, profile, roughness and other properties of the surfaces that would be in direct contact with the gas diffusion layer. Moreover, the scans were used to verify the exact placement of all the O-rings (diameter of 7 mm and total thickness of 1mm). Fig.9 depicts scans taken during the manufacturing and assembling processes. The scale in z-direction is set to $\gg 1$ to highlight the irregularities of the surface. The line-plots in Fig.9(c) and Fig.9(d) shows the precision needed for the different components to achieve perfect gas tightness. The observed irregularities were minimized by an inhouse sanding process to ensure good electrical and thermal contact between all segments and the gas diffusion layer.

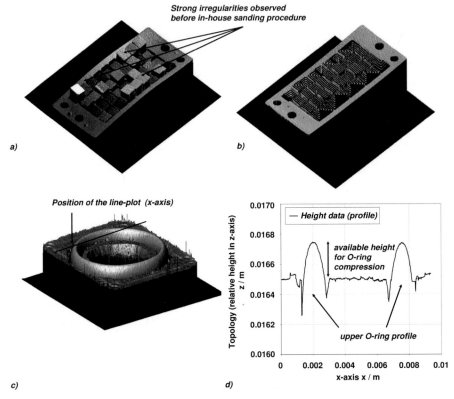

Fig.9. Surface scan of the segmented cathode side bipolar-plate after manufacturing and assembling and before sanding (x-y-z-plane) – without the milled flow-field (a), with the milled flow-field (b), and a detailed view of one of the 36 embedded O-rings (c). Topology of the embedded O-ring (line-plot along the x-axis), as indicated in c (d).

The cathode side of the cell was modified to allow for segmented measurements. In total, 36 bipolar-plate segments (4 x 9 = 36 segments of 1.347 cm² each, thickness of 3 mm), embedded in a polyetheretherketone insulation matrix (thickness of 10 mm), were used and contacted from the backside with gold-plated spring contact probes. For the cathode side aluminum endplate, 36 pinholes were drilled to insert 36 PT-100 resistive thermal devices (fully electrically insulated). The thickness of this endplate was increased from 12 mm to 18 mm to maintain structural stability during operation at high operating temperatures. An additional polyetheretherketone plate of 10 mm was used to fix 36 gold-plated spring contact probes and the 36 PT-100 devices to ensure good contact between the tips and the bipolar-plate segment. The anode side consisted of an aluminum endplate (thickness of 12 mm) equipped with two heating elements of 200 W, a polyetheretherketone insulation plate (thickness of 10 mm), a gold-plated copper current collector plate (thickness of 1 mm), and a bipolar-plate (thickness of 3 mm). Fig.10 shows the segmented cathode side bipolar-plates and the nonsegmented anode side bipolar-plates after being used in the test cell.

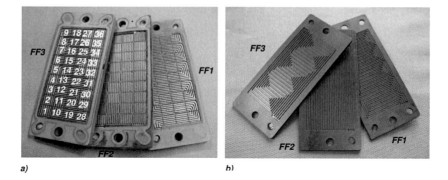

a)   b)

Fig.10. Segmented cathode side bipolar-plates after being used for all measurements. The number of each segment is given (as a mirrored representation because the cathode side is facing up in the experimental set-up) (a). Nonsegmented anode side bipolar-plates used for all measurements (b).

A commercially available PBI/H_3PO_4 sol-gel membrane electrode assembly (BASF Celtec®-P 1000 MEA) [38] with an active area of 48.5 cm^2 was used for all measurements (dimensions of approximately 10 cm x 5 cm). The cathode side reaction layer consisted of Vulcan XC-72 supported platinum-alloy with a loading of 0.75 mg cm^{-2}. The anode contained a Vulcan XC-72 supported platinum catalyst with a loading of 1 mg cm^{-2}.

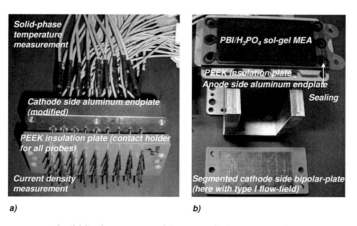

a)   b)

Fig.11. Components used within the segmented HTPEM fuel cell. Modified cathode side aluminum endplate with 36 PT-100 RTDs and 36 gold-plated spring contact probes (a). Segmented cathode side bipolar-plate, anode side endplate and polyetheretherketone insulation plate, and the membrane electrode assembly with the sealing (b).

Fig.11 shows all of the components used within the segmented HTPEM fuel cell including the custom made sealings made out of FKM material and the membrane electrode assembly. Perfect cell gas tightness was achieved by applying 2 Nm of torque to each of the 6 screws, leading to a compression of 8-11 bar on the gas diffusion layer surface at room temperature and pressure. The very same compression was applied to the nonsegmented HTPEM fuel cell. Finally, the entire assembly was thermally insulated against the environment to minimize heat loss and ensure accurate measurements.

6.3 Measuring the solid-phase temperature distribution

The solid-phase temperature distribution was measured using 36 PT-100 class A 4-wire resistance temperature detectors (RTDs) connected to a multiplexer from Agilent, U.S.A. [213] that sequentially scanned all of the channels with a resolution of 16 bits (Fig.12). The overall accuracy of the solid-phase temperature measurements, the following must be noted. The overall thermal conductivity of the bipolar-plate material is 20 W m^{-1} K^{-1} (through-plane) [138], whereas the in-plane thermal conductivity is assumed to be 10 times higher. The in-plane heat conduction within the bipolar-plate is influenced by the segmentation because the thermal conductivity of the insulation plate is in the range of 0.43 W m^{-1} K^{-1}. From this perspective, it is expected that the nonsegmented solid-phase temperature distribution will be slightly smoother than the reported distributions and that some accuracy will be lost due to segmentation. Nevertheless, the gas diffusion layer is not segmented, and conductive heat transport in the in-plane direction is guaranteed, even though the thermal in-plane and through-plane conductivities of the gas diffusion layer material are lower than those of the bipolar-plate material.

6.4 Measuring the current density distribution

The segmented cathode side bipolar-plate along with the 36 gold-plated spring contact probes, was used to measure the current density distribution. These probes were directly connected to a high-precision shunt resistance network from Bader, Germany (200 mΩ, ±10 ppm $°C^{-1}$) [214]. The shunt resistance network was analyzed prior to its use. All measurements were performed at room temperature and pressure. First, the values of the 36 shunts were individually verified while drawing a relatively high current through them. A mean value of 200.955 mΩ was confirmed at 21°C. Next, the resistance values from the load connector to the different channel connectors were examined, and values between 241 mΩ and 244 mΩ were measured. Once connected, the total resistance of the shunt resistance network, including all wiring and connectors, was measured to be 6.742 mΩ. The resulting additional voltage losses led to a slightly lower cell performance compared to the nonsegmented cell. When drawing current from the cell, the voltage drop over the 36 shunts was sequentially scanned with the same resolution of 16 bits (Fig.12). The highest load current per channel was limited to 1.388

A (resulting in a total load current of 50 A). The current density distribution may be slightly influenced by the segmentation of the bipolar-plate, directly connected to the nonsegmented gas diffusion layer, which has a higher electrical conductivity, especially in the in-plane direction.

6.5 Segmented EIS measurements

6.5.1 The simplified set-up for segmented EIS measurements

The segmented HTPEM fuel cell offered the possibility to perform segmented EIS measurements (sequentially scanned) to obtain a better understanding of the fuel cell operation. By combining these measurements with the solid-phase temperature and current density measurements it is possible to pinpoint the limiting factors under various operating conditions. Fig.12 depicts a simplified diagram of the experimental set-up. The hardware consisted of a high-precision electrochemical workstation, IM6eX (IM6 EPC-42) from Zahner Elektrik, Germany [56] and the electric load EL300. To choose an appropriate excitation signal, dummy measurements were performed for different signals (1 mV, 5 mV, 10 mV signals and 200 mA, 500 mA, and 1,000 mA signals), and the spectra were analyzed. No noticeable difference was observed, and thus, all of the spectra were recorded with 200 mA and 500 mA. Depending on the operating conditions, the modulation frequency ranged from 20 mHz up to 10 kHz. Experimental artifacts were minimized using twisted sense wires, short cable lengths, and the best possible cable connectors.

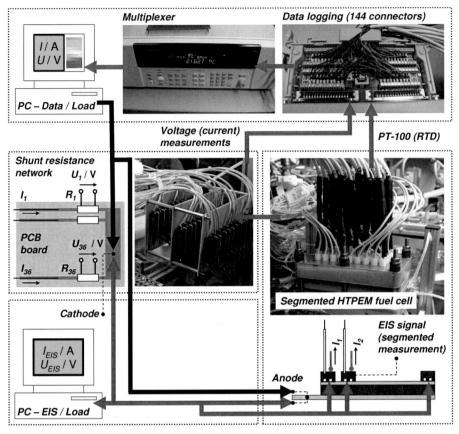

Fig.12. Simplified experimental set-up for performing segmented measurements.

6.5.2 Note on the experimental set-up

Due to the absence of hardware for simultaneous (parallel scanned) measurements, the segmented EIS measurements were performed sequentially rather than simultaneously. This drawback increased the recording times, and the EIS measurements were only performed at selected segments as a direct consequence. Another drawback was that only one segment was excited at a time, which could lead to erroneous interpretations associated with drift or transients between measurement cycles [206]. Hakenjos et al. [168] and Gerteisen et al. [201] discussed the benefits of multichannel characterization systems for segmented measurements in great detail and included voltage controls for each segment to minimize the voltage difference between adjacent segments and to eliminate current leaks along the highly conductive gas diffusion layer. However, the drawbacks of such a complex set-up were discussed by Reshetenko et al. [206]. A relatively simple segmented cell design minimizes the impact of the set-up (e.g., removal of the uniform voltage control limitation). Additionally, the use of standard testing procedures and the flexibility associated with the use of a single load are beneficial. In this work, for the current density measurements, the hardware was expected to slightly influence the current density distribution. Due to the nonsegmented membrane electrode assembly, lateral currents flowing between adjacent segments may disturb the current density measurement itself.

6.6 Equivalent circuit modeling

Various equivalent circuit models were tested, and it was found that the Randles model fit the local spectra well when the HTPEM fuel cell was operating at reference operating conditions. Nevertheless, this model was not able to accurately reproduce the impedance spectra for low stoichiometric flow rates and when operating the cell with CO enriched hydrogen. In this work, the equivalent circuit model shown in Fig.13 was used. This model was taken from the SIM software [56] model library and was modified accordingly [215]. This model was able to simulate the impedance spectra over a wide range of operating conditions for all three types of flow-fields.

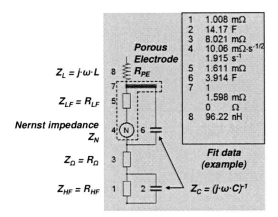

Fig.13. Equivalent circuit model of the HTPEM fuel cell (print screen from the SIM software).

The total ohmic resistance of the cell was described using a resistive element. An inductive element was used to represent the wiring and cables in the set-up. The capacitive element has the same features as an electronic capacitor and accounts for the double layer capacity. The high frequency loop within the local spectra was represented by a high frequency resistor capacitor pair and was associated with the charge transfer resistance (that is, the total high frequency resistance $R_{HF,tot}$). The most important contribution generally comes from the sluggish oxygen reduction reaction with a minor contribution from the hydrogen oxidation reaction and capacitive charge storage in the porous media [192]. The low frequency loop in the local spectra was related to the gas transfer resistance, which accounts for convective transport behaviour in the flow-field and the transfer of gases through the porous media towards the reaction layer [192]. The cathode side half-cell mainly contributes to this impedance, whereas the anode side only has a minor contribution to it. In this work, the gas transfer resistance is related to the total low frequency resistance and is calculated with the following equation.

$$R_{LF,tot} = R_{LF} + R_{PE} + R_N \tag{96}$$

As mentioned in [192], the gas transfer was added to model the second loop, appearing, for example, at low stoichiometric flow rates. It will act as a part of the charge transfer resistance when no second loop is present. The Nernst resistance accounted for the contribution of the diffusion process as part of the gas transfer reaction (equivalent to the general Warburg impedance) [56]. The Nernst impedance is calculated with the following equation.

$$Z_N = \frac{W}{\sqrt{j \cdot \omega}} \cdot \tanh\left(\sqrt{\frac{j \cdot \omega}{k_N}}\right) \tag{97}$$

The Nyquist plot exhibits the same shape as the (special) Warburg impedance diagram in the high frequency part and is similar to a resistor capacitor pair in the low frequency part [56]. Due to possible interactions with the Nernst impedance, a capacitive element was used instead of a constant phase element within the equivalent circuit. The porous electrode described the simulated impedance of a system with homogenous pores. Its contribution to the impedance was minor, but this component was included in the equivalent circuit model to obtain a smaller fitting error.

6.7 Analysis of the EIS data

The software package Thales 4.12 USB was used to evaluate the recorded data, and the SIM software was used to fit the experimental data to an equivalent circuit. One feature of this software was the stringent error treatment through all of the data sampling and processing steps. The measurement algorithm provides uncertainty estimates for each impedance sample [56]. For the fitting procedure itself, the recorded EIS data were first loaded into the software and then automatically smoothed. Next, the Z-HIT transform recreated the impedance data from the phase and frequency data to evaluate the steady-state and to treat any steady-state violations in the experimental data [56].

$$\ln|H(\omega_0)| \approx c + \frac{2}{\pi} \cdot \int_{\omega_s}^{\omega_0} \varphi(\omega) \cdot d \ln \omega + \gamma \cdot \frac{d\varphi(\omega_0)}{d \ln \omega} \tag{98}$$

The fitting samples were then selected exclusively from the Z-HIT data, primarily in the frequency range of 20-50 mHz to 10 kHz. Before calling the fitter, the values of the elements in the equivalent circuit were tuned to provide the best possible initial conditions. The SIM software utilized a complex nonlinear regression least square fitting algorithm. This fitting algorithm was applied in an accurate and stable manner and was adapted to the individual parameter behaviour for the impedance elements to optimize the model parameters and minimize the deviation between the model transfer function and data set [56]. After returning a solution, all fitting errors (overall error, impedance error, and phase angle error) were examined, and the returned parameters were verified. The reported uncertainty and significance of all of the fitted EIS model parameters were also examined.

6.8 Operating the nonsegmented and segmented HTPEM fuel cell

Both cells were operated in a dedicated teststand for fuel cell modeling validation. The teststand was equipped with all of the necessary hardware (e.g., differential pressure transducers, mass flow controllers, valves) and was controlled (e.g., implemented PID-controller) using LabVIEW™ from National Instruments, U.S.A. [216]. During all measurements, the data were recorded (field-points with a resolution of 16 bits) and displayed in a graphical user interface. The set solid-phase temperature was controlled with four type-K thermocouples placed at different locations on the HTPEM fuel cell. The fluid-(gas)-phase temperature was measured close to the gas inlets and outlets with additional four type-K thermocouples. All temperatures and segmented data were continuously evaluated during heating up, at no-load operating conditions and at load operating conditions. The nonsegmented HTPEM fuel cell was heated-up using 4 heating elements of 200 W (2 at the anode side and 2 at the cathode side) integrated in the aluminum endplates. This approach guaranteed relatively short heat-up times. Starting at 21°C, the set solid-phase temperature of 110°C was reached after less than 20 minutes. Next, power was drawn from the cell. A different heat-up procedure was used for the segmented HTPEM fuel cell. A novel feature of it is that the membrane electrode assembly and both bipolar-plates were sandwiched between two polyetheretherketone plates as can be seen from Fig.11. Thus, the cell was designed in a way to minimize the influence of the 2 heating elements of 200 W at the anode side of the cell on the solid-phase temperature distribution. This influence can be neglected, and the slight heterogeneities in the measured distributions were subject to contact variations between the tip of the PT-100 RTD and the segments. However, the fact that the membrane electrode assembly and both bipolar-plates were sandwiched between two polyetheretherketone plates led to long heat-up times. It took about 90 minutes until a mean solid-phase temperature of 100-110°C was reached. Once the desired set solid-phase temperature was reached, it remained stable. The observed solid-phase temperature evolution demonstrated the effectiveness of the polyetheretherketone plates that sandwiched both bipolar-plates and the membrane electrode assembly. Fig.14 depicts the segmented

HTPEM fuel cell and the nonsegmented HTPEM fuel cell (with thermal insulation completely removed) during operation at the ZBT laboratory. In both set-ups, the cathode side was always facing up.

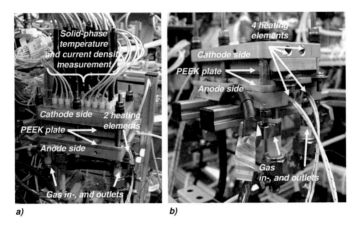

Fig.14. HTPEM fuel cells mounted within the dedicated teststand. Segmented HTPEM fuel cell (a). Nonsegmented HTPEM fuel cell (b).

Once a stable mean solid-phase temperature was observed at the cathode side, and after an additional waiting period of 10 minutes, gases were fed into the cell for two minutes and the load current increased with a rate of 1 A per minute. The cell was finally operated at the set load current and set solid-phase temperature, and all measurements were taken after 30 minutes of waiting time. When drawing power from the cell, the additional heating of the cathode side was exclusively due to the heat produced by the exothermal electrochemical reactions. After all measurements were collected for one type of flow-field, the entire cell was cooled down and disconnected from the teststand. The bipolar-plates were exchanged while maintaining the same membrane electrode assembly and same sealings (if possible) for one set of tests. After reassembling the set-up, exactly the same amount of torque was applied to guarantee the above mentioned gas tightness and very similar contact resistances.

7. Characterizing the three types of flow-fields

7.1 Fluid-flow distribution and pressure drop

7.1.1 Type I flow-field

It is expected that the distribution of the fluid-flow will greatly influence the overall HTPEM fuel cell performance. In fuel cells, the fluid flow is generally analyzed using transparent cells and particle image velocimetry (PIV-measurements), see, e.g., [91]. In this work, CFD modeling and simulation is used to examine the fluid-flow distribution and pressure drop. Fig.15 returns the fluid-flow distribution for type I flow-field and the pressure drop for an air flow rate of 1 l min^{-1}.

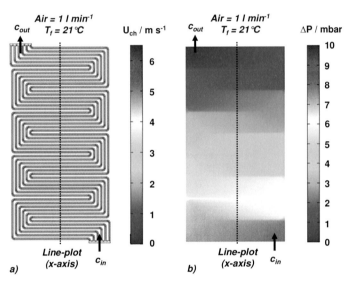

Fig.15. Simulation results for the cathode side (slice plot in x-y-plane). Fluid-flow distribution within the gas channels ($z = 500$ µm) (a) and pressure drop within the gas diffusion layer ($z = -1$ µm) (b).

The fluid-flow distribution is homogeneous throughout the flow-field, and the air strictly follows the gas channels. The highest velocity of 6.385 m s^{-1} occurs within the middle of the gas channel (Reynolds number of 383). The simulations indicate that the mean gas channel velocity is 2.344 m s^{-1} and the mean velocity within the porous media is 0.0018 m s^{-1}. The pressure drop is 8.079 mbar. As the air flow rate is increased to 2 l min^{-1}, the highest gas channel velocity increases to 12.886 m s^{-1}, with a mean value of 4.615 m s^{-1} (Reynolds number of 807). The mean velocity within the porous media is then 0.0043 m s^{-1} and the pressure drop increases to 19.210 mbar. Fig.16 depicts the fluid-flow distribution as line-plot. In Fig.16(a), minor fluctuations between the 6 gas channels can be observed.

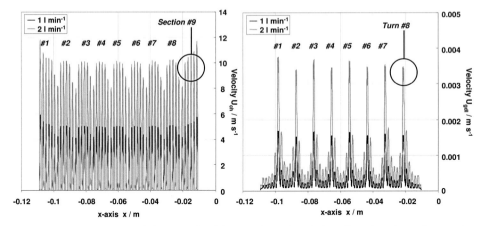

Fig.16. Simulation results of the cathode side gas channel velocity (line-plot along the x-axis as indicated in Fig.15 ($z = 500$ μm)) (a) and cathode side gas diffusion layer velocity (line-plot along the x-axis as indicated in Fig.15 ($z = -1$ μm)) (b).

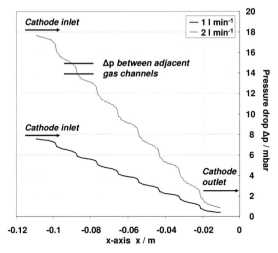

Fig.17. Simulation results of the cathode side pressure drop (line-plot along the x-axis as indicated in Fig.15 ($z = -1$ μm)).

The pressure drop in Fig.17 is represented as line-plot. The distribution at the $180°$ turns is of great importance. In this area, a pressure drop exists between adjacent channels and some gas is forced to flow from channel to channel via the gas diffusion layer. The velocity peaks in these regions can be seen in Fig.16(b),

7.1.2 Type II flow-field

As can be seen from Fig.18, for type II flow-field, the fluid-flow distribution is inhomogeneous throughout the flow-field. For an air flow rate of 1 l min^{-1}, the mean gas channel velocity is 0.835 m s^{-1} (Reynolds number of 868), and the mean porous media velocity is 0.0014 m s^{-1}. The highest velocity of 12.886 m s^{-1} occurs in the lateral gas channels close to the inlet and outlet. The pressure drop is 3.249 mbar. As the air flow rate is increased to 2 l min^{-1}, the mean gas channel velocity is 1.684 m s^{-1} and the highest velocity is 33.950 m s^{-1} (Reynolds number of 1,718). Within the porous media, the mean velocity increases to 0.0033 m s^{-1}, and the pressure drop reaches 9.240 mbar.

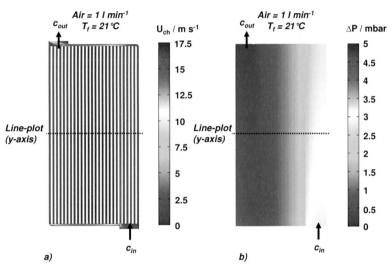

Fig.18. Simulation results for the cathode side (slice plot in x-y-plane). Fluid-flow distribution within the gas channels ($z = 500$ µm) (a) and pressure drop within the gas diffusion layer ($z = -1$ µm) (b).

Fig.19 shows the velocity distribution in more detail as line-plot. The highest gas channel velocities of 2-3 m s^{-1} at 1 l min^{-1} and of 4-5 m s^{-1} at 2 l min^{-1} are observed within the gas channels that are in direct contact with the gas inlet and outlet, that is, the 5 outermost gas channels. In the middle of the flow-field, very low gas channel velocities are observed. This poor distribution is a result of the fact that only one lateral gas channel connects all of the 26 parallel gas channels, as discussed in [171,182].

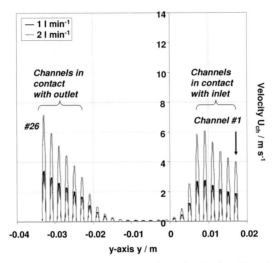

Fig.19. Simulation results of the cathode side gas channel velocity (line-plot along the *y*-axis as indicated in Fig.18 ($z = 500$ µm)).

7.1.3 Type III flow-field

For type III flow-field, the fluid-flow distribution is more homogeneous than it is the case for type II flow-field (Fig.20). For an air flow rate of 1 l min^{-1}, a gas channel velocity of 6.282 m s^{-1} is observed within the middle of the 6 gas channels that are connected to the inlet and outlet. The mean gas channel velocity is 1.232 m s^{-1} (Reynolds number of 295), and the mean velocity within the porous media is 0.001 m s^{-1}. The pressure drop reaches 2.290 mbar. When increasing the air flow rate to 2 l min^{-1}, the mean gas channel velocity is 2.464 m s^{-1} and the highest velocity is 12.580 m s^{-1} (Reynolds number 631). The pressure drop is now 6.012 mbar. The gas channel velocities within the three parallel regions of the flow-field are approximately three times lower because 2 x 3 parallel channels are used herein. Within the porous media, the mean gas velocity is 0.0020 m s^{-1}. Fig.21 and Fig.22 provide the fine details of the above discussed aspects as line-plot.

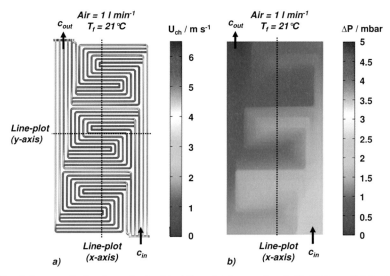

Fig.20. Simulation results for the cathode side (slice plot in x-y-plane). Fluid-flow distribution within the gas channels ($z = 500$ μm) (a) and pressure drop within the gas diffusion layer ($z = -1$ μm) (b).

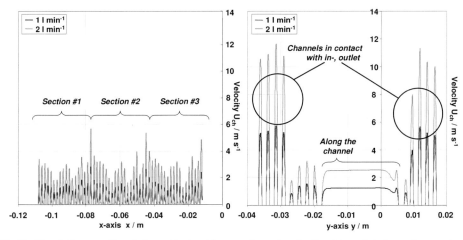

Fig.21. Simulation results of the cathode side gas channel velocity. Line-plot along the x axis as indicated in Fig.20 ($z = 500$ μm) (a) and line-plot along the y-axis as indicated in Fig.20 ($z = 500$ μm) (b).

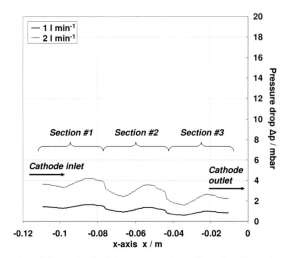

Fig.22. Simulation results of the cathode side pressure drop (line-plot along the *x*-axis as indicated in Fig.20 ($z = -1$ μm)).

The results demonstrated that type I flow-field returns the highest mean gas channel velocity and highest mean porous media velocity. It is remarkable that the pressure drop for type III flow-field is slightly lower than that for type II flow-field. Type I flow-field exhibits the highest pressure drop. The same pressure drop measurements were performed for the three types of flow-fields at no-load operating conditions. Generally spoken, the measured pressure drops were higher than the simulated ones. The simulations account for the net flow-field pressure drop from inlet to outlet but do not account for any additional pressure drops due to feeders, manifolds, connectors, pipes, or peripheral components. In fact, a more complex fluid-flow behaviour will likely exist within these regions (e.g., recirculation zones). Moreover, the depth of the modeled flow-fields is exactly 1 mm and does not account for a partial intrusion of the gas diffusion layer into the gas channel. Several dummy simulations were performed for type I flow-field and it was found that a 10% reduction of the cross-sectional area of the gas channel (*y-z*-plane) leads to a pressure drop of 12.710 mbar (1 l min^{-1}). Using analytical calculations, a pressure drop of 14.740 mbar was determined in this case. These values agree with the simulated pressure drop.

7.2 Impedance measurements at no-load operating conditions

Similar to the high-precision shunt resistance network, the nonsegmented and segmented HTPEM fuel cells were analyzed before operation. For the impedance measurements, the entire fuel cell is compressed by applying the same amount of torque to each of the 6 screws, leading to constant compression over the entire gas diffusion layer surface. The EIS measurement device is connected to the cell, and impedance measurements were performed at a frequency of 1 kHz with an EIS signal of 10 mV (potentiostatic mode). The mean resistance from the tip of the uncompressed gold-plated spring contact probe to the connector of the shunt resistance network is 25.85 mΩ.

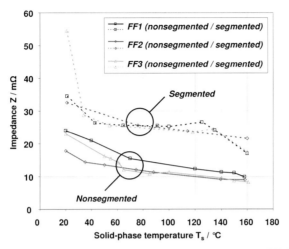

Fig.23. Impedance measurements of the nonsegmented and segmented HTPEM fuel cell equipped with the three types of flow-fields.

The impedance of the different set-ups is also measured at different temperatures, as can be seen from Fig.23. A decrease in impedance is observed with increasing temperature. At 21°C, impedances of 34.50 mΩ (set-up with type I flow-field), 32.54 mΩ (set-up with type II flow-field), and 54.50 mΩ (set-up with type III flow-field) are measured. At 160°C, the impedance decreases to 16.90 mΩ, 22.55 mΩ, and 17.50 mΩ respectively. Due to the segmentation and layout of the entire assembly, these values are higher than those for the nonsegmented cell. For the set-up with type I flow-field, 9.90 mΩ are measured, for the set-up with type II flow-field 9.01 mΩ and 8.29 mΩ for the set-up with type III flow-field are measured at 160°C. These values are in accordance with those published in [41].

7.3 Overall performance at reference operating conditions

The reference operating conditions summarized in Table 12 were used to compare the performance of the HTPEM fuel cells with the three types of flow-fields. The same operating conditions were employed for the boundary conditions in the modeling and simulation.

Table 12

Reference operating conditions.

Flow-field configuration	Conter-flow / co-flow	
T_∞ (ambient temperature (°C))	21°C	
$T_{s,0}$ (solid-phase temperature (°C))	160°C	
	Anode side	Cathode side
Gas	Hydrogen	Air
Flow rate	Stoichiometry of 1.3	Stoichiometry of 2.5
$T_{f,0}$ (gas inlet temperature (°C))	21	21
Gas humidity (%rh)	0.10 (virtually dry)	0.50 (virtually dry)
Outlet pressure (Pa)	$1.01325 \cdot 10^5$ (ambient)	$1.01325 \cdot 10^5$ (ambient)

7.3.1 Type I flow-field

Fig.24 displays the performance curves obtained as the cell was operated in co-flow and counter-flow configuration at reference operating conditions. The performance is very similar to the results published in [41]. The open circuit voltage is measured to be approximately 0.869 V, which is lower than the theoretical voltage of approximately 1.140 V at 160°C [41]. At a low current density, a sharp voltage drop is observed.

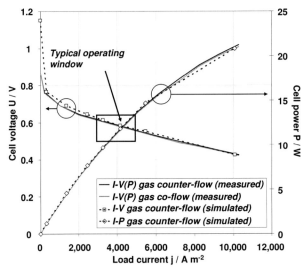

Fig.24. Simulated and measured performance curves when operating the HTPEM fuel cell in co-flow and counter-flow configuration at reference operating conditions.

For a high current density, a relatively low and almost constant voltage drop of 0.027 V per 5 A load current increase is observed (an increase of 5 A load current corresponds to a current density increase of 1,030 A m^{-2}). The highest load current is set to 50 A, corresponding to a current density of 10,310 A m^{-2}. For a typical load current of 20 A (4,123 A m^{-2}), the cell voltage in counter-flow configuration is 0.583 V and 0.580 V in co-flow configuration.

7.3.2 Type II flow-field

The overall performance for this type of flow-field is shown in Fig.25. The cell exhibits a difference of approximately 10 mV between the co-flow and counter-flow configuration. The measured curves display typical HTPEM fuel cell operating behaviour. The open circuit voltage is measured to be approximately 0.840 V, which is lower than the theoretical voltage at 160°C. Similar to type I flow-field, at a low current density, a sharp voltage drop is observed. For a high current density, an almost constant voltage drop of 0.055 V per 5 A load current increase is seen (an increase of 5 A load current is equivalent to a current density increase of 1,030 A m^{-2}). The highest load current is set to 35 A, corresponding to a current density of 7,216 A m^{-2}. For a typical load current of 20 A (4,123 A m^{-2}), the cell voltage in counter-flow configuration was 0.503 V and 0.492 V in co-flow configuration.

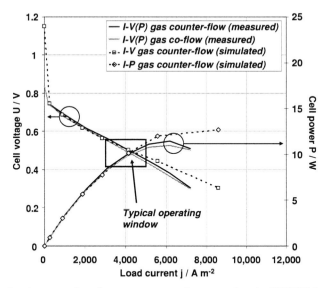

Fig.25. Simulated and measured performance curves when operating the HTPEM fuel cell in co-flow and counter-flow configuration at reference operating conditions.

7.3.3 Type III flow-field

The performance of type III flow-field is illustrated in Fig.26. A difference between the co-flow and counter-flow configuration is a few microvolts. The open circuit voltage is measured to be approximately 0.850 V. At a low current density, a sharp voltage drop is observed. For a high current density, an almost constant voltage drop of 0.028 V per 5 A load current increase is observed (an increase of 5 A load current is equivalent to a current density increase of 1,030 A m^{-2}). The highest load current is set to 50 A. For a typical load current of 20 A (4,123 A m^{-2}), the cell voltage in counter-flow configuration is 0.574 V and 0.570 V in co-flow configuration.

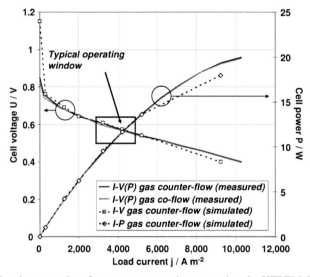

Fig.26. Simulated and measured performance curves when operating the HTPEM fuel cell in co-flow and counter-flow configuration at reference operating conditions.

The simulation results for counter-flow configuration are shown in Fig.24, Fig.25, and Fig.26. A good overall agreement is found for type I and type III flow-fields in counter-flow configuration (identical for co-flow configuration). For type II flow-field, the simulation results agree up to a current density of 4,500 A m^{-2} but fail to match the exact data at a high current density. Of the three types of flow-fields, type I flow-field displayed the best performance, whereas type II flow-field displayed the worst performance. Type III exhibited the second best performance, only slightly below the performance of type I flow-field. For all flow-fields, the performance of the cell operating in co-flow configuration was a few microvolts lower than that for counter-flow configuration.

8. Segmented solid-phase temperature and current density measurements

8.1 Solid-phase temperature distribution at no-load operating conditions

8.1.1 Type I flow-field

Before the measurements are taken, the set-up is heated-up until the set solid-phase temperature of 140°C is reached. Fig.27(a) displays a uniform solid-phase temperature distribution. For the results shown in Fig.27(b), dry air with a fluid-(gas)-phase temperature of 21°C is fed at an air flow rate of 1 l min^{-1}. The air that enters the cell influences the solid-phase temperature distribution, and lower values are observed close to the cathode inlet. Heat is absorbed from the air stream and transported towards the cathode outlet. Fig.27(c) depicts the situation when increasing the air flow rate from 1 to 1.5 l min^{-1}. The homogeneous fluid-flow distribution influences the solid-phase temperature distribution over a large area close to the cathode inlet. A slightly higher overall solid-phase temperature is measured towards the cathode outlet because heat accumulates throughout the total length of the flow-field. In the next step, the air flow rate was increased from 1.5 to 2 l min^{-1} (Fig.27(d)). In this case, the air entering the cell significantly influences the solid-phase temperature distribution over 35-40% of the membrane electrode assembly area. A higher overall solid-phase temperature is observed towards the cathode outlet. The shape of type I flow-field dictates the temperature evolution from cathode inlet to cathode outlet. Table 13 summarizes the simulated and measured results.

Table 13

Simulated and measured solid-phase temperature at no-load operating conditions.

Flow rate	Mean	Minimum	Maximum	Difference
l min^{-1}	°C	°C (%)	°C (%)	°C (%)
0	139.52	137.59 (98.62)	140.97 (101.04)	3.38 (2.42)
1	140.15	137.09 (97.81)	142.47 (101.65)	5.38 (3.83)
1.5	139.61	133.42 (95.56)	143.09 (102.50)	9.67 (6.92)
2	138.58	129.33 (93.32)	142.34 (102.71)	13.01 (9.38)
1 (simulated)	139.03	137.48 (98.88)	139.64 (100.44)	2.16 (1.55)
2 (simulated)	138.65	136.97 (98.78)	139.53 (100.63)	2.56 (1.84)

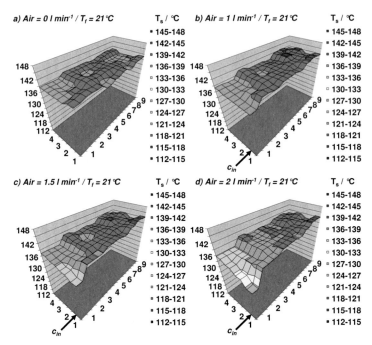

Fig.27. Measured solid-phase temperature distribution. Set solid-phase temperature 140°C and air inlet temperature 21°C. After heating up (a), air flow rate 1 l min⁻¹ (b), 1.5 l min⁻¹ (c), and 2 l min⁻¹ (d).

Fig.28 presents the simulated results for the solid-phase temperature distribution for an air flow rate of 1 l min⁻¹ (a) and 2 l min⁻¹ (b). The influence of the air entering the cell can clearly be seen at the cathode inlet. When comparing the simulation results with the measured result shown in Fig.27, it is seen that the same overall behaviour is found. The cooling effect shown in Fig.28 close to the cathode inlet is much less pronounced compared to the measurements. One reason for this discrepancy may be that the laminar inflow condition was not able to account for the complex fluid-flow distribution at the cathode inlet, possibly leading to higher local heat transfer coefficients. Alternatively, the segmentation of the bipolar-plate itself may be the cause of this difference.

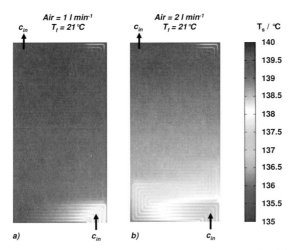

Fig.28. Simulated solid-phase temperature distribution within the cathode side gas diffusion layer (slice plot in *x-y*-plane, *z* = -1 μm). Set solid-phase temperature 140°C and air inlet temperature 21°C. Air flow rate 1 1 min^{-1} (a) and 2 1 min^{-1} (b).

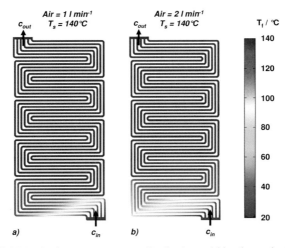

Fig.29. Simulated fluid-(gas)-phase temperature distribution within the cathode side gas channels (slice plot in *x-y*-plane, *z* = 500 μm). Set solid-phase temperature 140°C and air inlet temperature 21°C. Air flow rate 1 1 min^{-1} (a) and 2 1 min^{-1} (b).

From Fig.29 it can be seen that the fluid-(gas)-phase temperature increases as the air flows along the channel. The air leaves the cell at the cathode outlet with a temperature of 139.50°C at an air flow rate of 1 1 min^{-1}. In this case, 1.158 W are transferred through the gas channel (or bipolar-plate) walls. Within the gas diffusion layer 0.434 W are picked up by the air, and within the reaction layer is 0.047 W. At an air flow rate of 2 1 min^{-1}, the air outlet temperature is 139.31°C. In this case, 2.1 W are transferred from the solid-phase to the fluid-(gas)-phase through the gas channel wall, 0.807 W are transferred within the gas diffusion layer, and 0.088 W are transferred within the reaction layer.

8.1.2 Type II flow-field

The results for type II flow-field are depicted in Fig.30. Before the air is fed into the cell, the solid-phase temperature distribution is quite uniform. When increasing the air flow rate, the distribution becomes more inhomogeneous. Heat accumulates towards the cathode outlet, especially in the region in which the lowest gas channel velocity is observed, corresponding to the region of the innermost gas channels, as depicted in Fig.19. In addition, the solid-phase temperature distribution on the right side over the membrane electrode assembly area is somewhat lower than that on the left side because the 5 outermost channels are in direct contact with the cathode inlet and cathode outlet. Fig.18 illustrates that a higher gas channel velocity exists in these channels, resulting in more effective cooling within these gas channels. Thus, the shape of type II flow-field dictates the solid-phase temperature distribution over the membrane electrode assembly area. Table 14 summarizes the simulated and measured results.

Table 14

Simulated and measured solid-phase temperature at no-load operating conditions.

Flow rate	Mean	Minimum	Maximum	Difference
l min^{-1}	°C	°C (%)	°C (%)	°C (%)
0	139.33	136.78 (98.17)	141.56 (101.6)	4.78 (3.43)
1	138.21	135.51 (98.04)	140.72 (101.81)	5.21 (3.76)
2	138.97	135.07 (97.19)	142.72 (102.70)	7.65 (5.50)
2.5	138.75	131.34 (94.66)	145.44 (104.82)	14.10 (10.16)
1 (simulated)	139.46	135.94 (97.47)	139.85 (100.28)	3.91 (2.80)
2 (simulated)	138.81	135.24 (97.42)	139.63 (100.59)	4.39 (3.16)

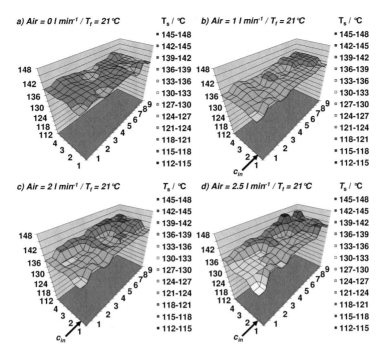

Fig.30. Measured solid-phase temperature distribution. Set solid-phase temperature 140°C and air inlet temperature 21°C. After heating up (a), air flow rate 1 l min^{-1} (b), 2 l min^{-1} (c), and 2.5 l min^{-1} (d).

Fig.31 displays the simulated results for the solid-phase temperature distribution for an air flow rate of 1 l min^{-1} (a) and 2 l min^{-1} (b). The simulated distribution is similar to the measurements. A lower solid-phase temperature is observed close to the cathode inlet, especially on the right side over the membrane electrode assembly area (outermost gas channels). The simulations also indicate that a slightly higher solid-phase temperature occurs in the middle region over the membrane electrode assembly area and towards the cathode outlet.

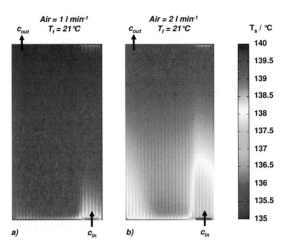

Fig.31. Simulated solid-phase temperature distribution within the cathode side gas diffusion layer (slice plot in *x-y*-plane, *z* = -1 μm). Set solid-phase temperature 140°C and air inlet temperature 21°C. Air flow rate 1 l min⁻¹ (a) and 2 l min⁻¹ (b).

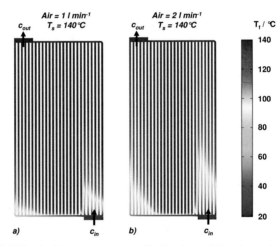

Fig.32. Simulated fluid-(gas)-phase temperature distribution within the cathode side gas channels (slice plot in *x-y*-plane, *z* = 500 μm). Set solid-phase temperature 140°C and air inlet temperature 21°C. Air flow rate 1 l min⁻¹ (a) and 2 l min⁻¹ (b).

The fluid-(gas)-phase temperature increases as the air flows along the channel towards the cathode outlet (Fig.32). The fluid-flow distribution shown in Fig.18 and Fig.19 clearly dictates the shape of the fluid-(gas)-phase temperature distribution. The air leaves the cell at the cathode outlet with a temperature of 139.76°C at an air flow rate 1 l min⁻¹. In this case, 1.203 W are transferred through the gas channel walls, 0.454 W are transferred within the gas diffusion layer and 0.053 W are transferred within the reaction layer. At an air flow rate of 2 l min^{-1} the cathode outlet temperature is 139.09°C. In this case, 2.398 W are transferred through the gas channel wall, 0.981 W are transferred within the gas diffusion layer, and 0.110 W are transferred within the reaction layer.

8.1.3 Type III flow-field

The influence of the air flow rate on the solid-phase temperature distribution for type III flow-field is depicted in Fig.33. At an air flow rate of 1 l min^{-1}, the distribution is only slightly influenced. The only difference is the somewhat higher solid-phase temperature in the middle region over the membrane electrode assembly area (Fig.33(b)). The influence of the air entering the cell becomes more visible at higher flow rates, as shown in Fig.33(c) and Fig.33(d). The mean solid-phase temperature decreases, whereas the highest overall solid-phase temperature is observed in the middle region over the membrane electrode assembly area and towards the cathode outlet. These results are in accordance with the fluid-flow distribution depicted in Fig.20 to Fig.22. The shape of type III flow-field dictated the solid-phase temperature distribution from cathode inlet to cathode outlet. Table 15 summarizes the simulated and measured results.

Table 15

Simulated and measured solid-phase temperature at no-load operating conditions.

Flow rate	Mean	Minimum	Maximum	Difference
l min^{-1}	°C	°C (%)	°C (%)	°C (%)
0	141.49	139.25 (98.41)	142.92 (101.01)	3.67 (2.59)
1	141.28	138.70 (98.17)	142.73 (101.03)	4.03 (2.85)
2	139.86	136.84 (97.83)	142.65 (101.99)	5.81 (4.15)
2.25	138.73	135.46 (97.64)	143.09 (103.15)	7.63 (5.5)
1 (simulated)	139.27	137.20 (98.51)	139.77 (100.36)	2.57 (1.84)
2 (simulated)	138.56	136.56 (98.55)	139.50 (100.68)	2.94 (2.12)

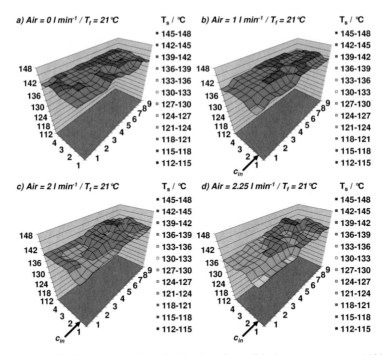

Fig.33. Measured solid-phase temperature distribution. Set solid-phase temperature 140°C and air inlet temperature 21°C. After heating up (a), air flow rate 1 l min^{-1} (b), 2 l min^{-1} (c), and 2.25 l min^{-1} (d).

Fig.34 presents the simulated results for the solid-phase temperature distribution for an air flow rate of 1 l min^{-1} (a) and 2 l min^{-1} (b). The influence of the air entering the cell can be seen at the cathode inlet. The solid-phase temperature continuously increases towards the cathode outlet. The overall behaviour is similar to the measured distribution. Fig.35 presents the fluid-(gas)-phase temperature distribution. The fluid-flow distribution dictates the shape of the temperature evolution from cathode inlet to cathode outlet. For an air flow rate of 1 l min^{-1}, 1.295 W are transferred through the gas channel walls, 0.475 W are transferred within the gas diffusion layer, and 0.052 W are transferred within the reaction layer. At an air flow rate of 2 l min^{-1}, 2.473 W are transferred through the gas channel wall, 0.965 W are transferred within the gas diffusion layer, and 0.106 W are transferred within the reaction layer. When comparing the reported results to each other, type I flow-field exhibits the greatest difference between the highest and lowest solid-phase temperatures. In most cases, the reported values are in the range of 94-104% of the set solid-phase temperature. The reported values for the total heat transfer between the fluid-(gas)-, and the solid-phase are in the range of 1- 3 W. For the three types of flow-fields the channel to wall area is approximately the same (Table 3). Thus, that the amount of heat that is transferred between the two phases is in the same order of magnitude for each case.

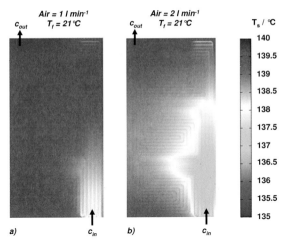

Fig.34. Simulated solid-phase temperature distribution within the cathode side gas diffusion layer (slice plot in x-y-plane, $z = -1$ μm). Set solid-phase temperature 140°C and air inlet temperature 21°C. Air flow rate 1 l min^{-1} (a) and 2 l min^{-1} (b).

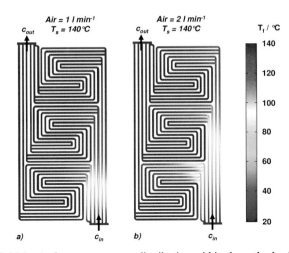

Fig.35. Simulated fluid-(gas)-phase temperature distribution within the cathode side gas channel (slice plot in x-y-plane, $z = 500$ μm). Set solid-phase temperature 140°C and air inlet temperature 21°C. Air flow rate 1 l min^{-1} (a) and 2 l min^{-1} (b).

The very same segmented solid-phase temperature measurements were performed at no-load operating conditions as hydrogen entered the cell with 21°C. No visible influence was observed for different hydrogen flow rates, and the gas was likely already heated-up to the set solid-phase temperature once it reached the inlet of the flow-field. The simulation results returned the same results.

8.2 Load operating conditions – Counter-flow configuration

8.2.1 Type I flow-field – Operation with hydrogen and air

Fig.36 presents the results obtained for the cell operating with hydrogen and air at constant flow rates of 0.2 l min^{-1} (anode) and 1 l min^{-1} (cathode). In Fig.36(a), a load current of 10 A is drawn from the cell. The highest current density is observed close to the cathode inlet. For these flow rates, the overall current density distribution is flat and decreases almost linearly towards the cathode outlet. The solid-phase temperature distribution is higher in the region of the cathode inlet and is primarily defined by the shape of the current density (Fig.36(b)). The air entering the cell at the cathode side influences the solid-phase temperature distribution because the constant flow rates are relatively high. The highest solid-phase temperature is measured in the region of the cathode inlet, whereas lower values are observed at the cathode inlet and towards the cathode outlet. When a load current of 15 A is drawn from the cell, the solid-phase temperature and current density gradient increase slightly (Fig.36(c) and Fig.36(d)). As the load current is further increased under a constant gas flow rate, a higher current density gradient is measured between the cathode inlet and cathode outlet, whereas the shape of the distribution remains the same (Fig.36(e)). Because more heat is produced, the mean solid-phase temperature distribution increases. The influence of the air entering the cell remains visible at the cathode inlet (Fig.36(f)). The solid-phase temperature distribution is fairly homogeneous for the given operating conditions. The distribution is defined by the shape of the flow-field and is overlapped by the heat released by the electrochemical reactions. As both stoichiometric flow rates are reduced while drawing 15 A load current, the current density distribution becomes extremely inhomogeneous (Fig.36(g)). The oxygen availability clearly defines the shape of the distribution. The mean solid-phase temperature increases, and a higher temperature is observed in the region of the highest current density. The influence of the fluid-(gas)-phase temperature close to the cathode inlet vanishes (Fig.36(h)).

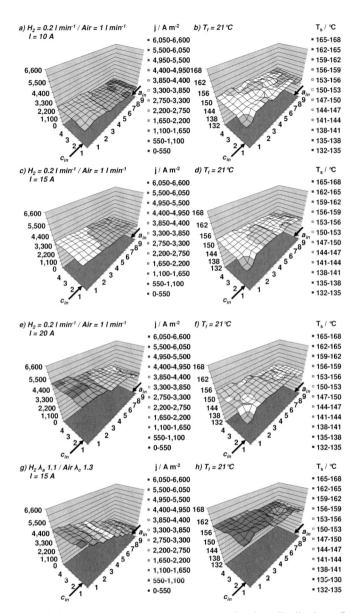

Fig.36. Measured solid-phase temperature and current density distribution. Set solid-phase temperature 150°C and gas inlet temperature 21°C. Constant flow rates for 10 A (a,b), 15 A (c,d), 20 A (e,f) and stoichiometric flow rates for 15 A load current (g,h).

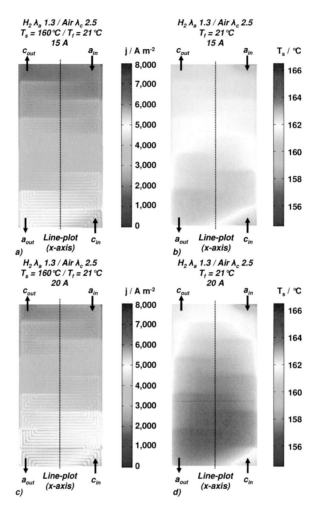

Fig.37. Simulated solid-phase temperature and current density distribution within the cathode side reaction layer (slice plot in x-y-plane, $z = -379$ μm). Set solid-phase temperature 160°C and gas inlet temperature 21°C. Stoichiometric flow rates for 15 A (a,b) and 20 A load current (c,d).

Selected simulation results are presented to support the measured solid-phase temperature and current density distribution. Fig.37 depicts the simulation results obtained at reference operating conditions while drawing 15 A and 20 A load current. These operating conditions are similar to the operating conditions in Fig.36(e) and Fig.36(f). Within the gas diffusion layer, the oxygen mass fraction is highest at the gas channel to porous media interface. From this interface it diffuses towards the reaction layer. Additionally, the oxygen availability decreases almost linearly along the gas channel (x-axis) due to ongoing electrochemical reactions. The distribution of the oxygen mass fraction strongly influences the current density distribution. Additionally, the distribution is also affected by the bipolar-plate or flow-field structure. At the cathode outlet, the oxygen mole fraction is calculated to be approximately 0.15 for the given stoichiometric flow rates and inlet molar fractions. Over the membrane electrode assembly area, higher current density values are observed under the ribs than under the channels of the bipolar-plate. Current density peaks occur in the regions in which the 180° bends are located. This result may be caused by the higher local velocity and reactant concentration within the reaction layer, as discussed above and noted in [82]. The shape of the simulated solid-phase temperature agrees with the measurements. Within the porous media, the distribution is slightly influenced by the gases entering the cell. The highest solid-phase temperature is seen in the region in which the highest current density occurs. For 15 A, the total heat produced within the cathode side reaction layer is 11.600 W. Of this value, the irreversible reaction heat comprises 68.11% (7.904 W), the reaction entropy comprises 31.4% (3.644 W), and the total Joule heating (resistive heating – electric and ionic) makes up the remaining 0.47% (0.055 W). The same trend occurs for 20 A, where 15.39 W are produced (irreversible reaction heat 68.91% (10.6 W), the reaction entropy 30.49% (4.61 W), resistive heating 0.58% (0.090 W)) and 25 A where 21.160 W are produced (irreversible reaction heat 70.07%, the reaction entropy 29.17%, resistive heating 0.74%).

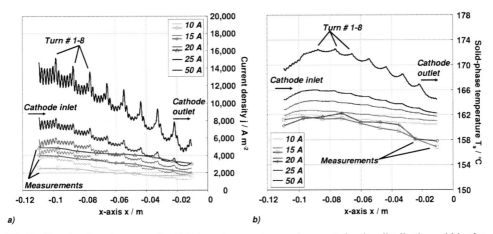

Fig.38. Simulated and measured solid-phase temperature and current density distribution within the cathode side reaction layer (line-plot along the *x*-axis as indicated in Fig.37 (z = -379 μm)). Set solid-phase temperature 160°C and gas inlet temperature 21°C. Stoichiometric flow rates for 10 A, 15 A, 20 A, 25A, and 50 A load current.

The line-plot shown in Fig.38 provides details about the current density distribution and highlights the aforementioned aspects. The overall behaviour of the current density distribution for 10 A, 15 A, and 20 A load current agrees with the simulations. Close to the cathode inlet, the current density under the land is considerably higher than that under the channel. The observed difference becomes less pronounced towards the cathode outlet for all load currents. Individual current density peaks at each

180° bend can be clearly observed. The above results indicate that the channel-to-land ratio should be seen as an optimization parameter for the flow-field layout used for HTPEM fuel cells, especially at the cathode side. As the load current increases, the difference between the lowest and highest current densities also increases. The somewhat lower current density values observed directly at the cathode inlet may be explained as follows. The water vapor mass fraction increases towards the cathode outlet and towards the reaction layer (z-axis). Higher values are observed under the land areas and close to the borders over the membrane electrode assembly area, especially towards the cathode outlet. These values increase with the load current. Additionally, a slightly higher water vapor mass fraction is observed under the land areas because it takes longer until this species reaches the gas channel. This explanation is reasonable because the water vapor must diffuse through the porous media towards the gas channel before it finally leaves the cell. At typical low humidity conditions in HTPEM fuel cells, the concentration of the phosphoric acid will vary greatly depending on the product water and temperature. Initially, the concentration is supposed to be 85wt.% at both sides of the cell. According to the simulation results, the concentration varies between 94.09wt.% (1 A load current) and 96.6wt.% (50 A load current) at the cathode side and from 106.54wt.% (1 A load current) to 102.07wt.% (50 A load current) at the anode side. Moreover, the simulations reveal that the phosphoric acid concentration varies in the region close to the inlets. The oxygen diffusion coefficient and oxygen solubility vary accordingly in the region close to the inlets and remain approximately constant elsewhere. Fig.38(b) depicts the simulation results for the solid-phase temperature distribution as line-plot. The highest solid-phase temperature occurs close to the region in which the highest current density is located. Due to the higher current density under the land area, the solid-phase temperature peaks accordingly. According to the simulations, at 15 A load current, the highest solid-phase temperature is 162.78°C, at 20 A it is 164.31°C, at 25 A it is 166.06°C, and at 50 A it is 172.56°C. That the solid-phase temperature is somewhat lower in the regions under the gas channels. This difference nearly vanishes towards the exit. The influence of the air entering the cell is noticed at the cathode inlet, and the measured values are slightly lower than the simulated ones.

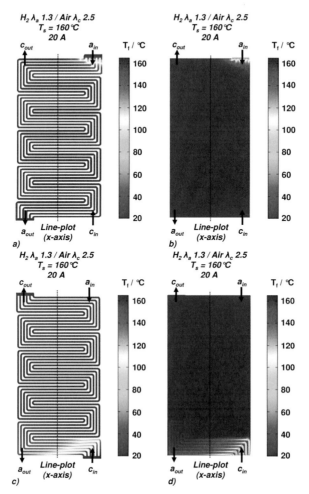

Fig.39. Simulated fluid-(gas)-phase temperature distribution within the anode (a,b) and cathode side gas channels and gas diffusion layer (c,d) (slice plot in *x-y*-plane, $z = -1547$ μm, $z = -1046$ μm, $z = 500$ μm, $z = -1$ μm). Set solid-phase temperature 160°C and gas inlet temperature 21°C. Stoichiometric flow rates for 20 A load current.

The fluid-(gas)-phase temperature for 20 A load current is shown in Fig.39(a) and Fig.39(b) (anode side) respectively Fig.39(c) and Fig.39(d) (cathode side). At the inlet of the gas channel, the fluid-(gas)-phase temperature is 21°C and increases rapidly until phase equilibrium is reached. The system takes slightly longer to achieve phase equilibrium within the gas channels than within the porous media. A close inspection of these figures reveals that the gas temperature increases slightly close to the bipolar-plate (gas channel) boundaries. Close to the gas channel to gas diffusion layer interface, the gases are already heated-up due to the fact that they diffuse through the porous media (the *z*-direction towards the reaction layer). At the anode side, hydrogen heats up quickly (high thermal conductivity and high heat capacity) and reaches the set solid-phase temperature shortly after entering the gas channels. As more current is drawn, more gas enters the cell due to stoichiometric flow rates, and thus, it takes longer for the gas to reach the set solid-phase temperature. More heat needs to be exchanged between the two phases, and the thermal equilibrium between the solid-, and the fluid-(gas)-phase shifts in the direction of the channel.

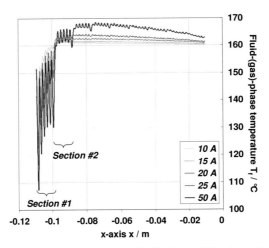

Fig.40. Simulated fluid-(gas)-phase temperature distribution within the cathode side gas diffusion layer. Set solid-phase temperature 160°C and gas inlet temperature 21°C (line-plot along the x-axis as indicated in Fig.39 ($z = -1$ μm)). Stoichiometric flow rates for 10 A, 15 A, 20 A, 25A, and 50 A load current.

Fig.40 highlights the fluid-(gas)-phase temperature along the x-axis close to the gas channel to gas diffusion layer interface at the cathode side ($z = -1$ μm) for different load currents. The temperature in the regions under the gas channels is somewhat lower than that under the land, especially for high load currents. In most cases, thermal equilibrium is reached after the first 180° bend of the flow-field. The fluid-(gas)-phase temperature is relatively smooth towards the outlets. The highest fluid-(gas)-phase temperature corresponds to the highest solid-phase temperature and increases for higher load currents.

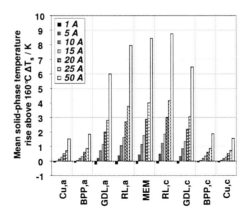

Fig.41. Simulated mean solid-phase temperature rise above the set solid-phase temperature within the different components. Set solid-phase temperature 160°C and gas inlet temperature 21°C. Stoichiometric flow rates for 1 A, 5 A, 10 A, 15 A, 20 A, 25A, and 50 A load current.

Fig.41 depicts the simulated mean solid-phase temperature rise above 160°C for all cell components. The simulation results reveal that a nearly uniform solid-phase temperature distribution exists within the gold-plated copper current collectors and in both bipolar-plates due to the high thermal conductivities of both materials compared to all other thermal conductivities within the set-up. Only a slightly higher solid-phase temperature is observed within these components when drawing higher load currents. The mean solid-phase temperature rise is more pronounced within the gas diffusion layer, reaction layer, and membrane. It is also noticed, that for a very low load currents of 1 A, the mean solid-phase temperature slightly falls because more heat is transferred out of the cell than is produced within the cell.

Table 16

Simulated heat transfer between the two phases / W

	Load current I / A			
	15	20	25	50
Anode side component				
Gas channel	0.248	0.339	0.419	0.817
GDL	0.248	0.346	0.725	0.894
RL	0.052	0.073	0.097	0.200
Cathode side component				
Gas channel	1.428	1.902	2.405	4.864
GDL	0.370	0.491	0.440	1.372
RL	0.050	0.069	0.115	0.226

Table 16 summarizes the simulated heat transfer between the two phases for selected load currents. Within the gas channels, the heat is absorbed by the gas and transported out of the cell. According to the simulations, more heat is transferred between the two phases at the cathode side than at the anode side. Table 17 summarizes the simulated and measured results.

Table 17

Overview of simulated and measured solid-phase temperature and current density distribution for various operating conditions. Set solid-phase temperature and set current density equals 100%.

	Solid-phase temperature T_s and current density j			
	Mean	Minimum	Maximum	Difference
	°C	°C (%)	°C (%)	°C (%)
	A m^{-2}	A m^{-2} (%)	A m^{-2} (%)	A m^{-2} (%)
Measurement				
H_2 0.2 l min^{-1}, Air 1 l min^{-1}, 10 A	150.97	147.70 (97.83)	153.50 (101.68)	5.80 (3.84)
	2,061	1,630 (79.47)	2,545 (124.09)	915 (44.61)
H_2 0.2 l min^{-1}, Air 1 l min^{-1}, 15 A	151.40	145.60 (96.16)	153.98 (101.70)	8.38 (5.53)
	3,092	2,251 (72.73)	3,725 (120.37)	1,474 (47.62)
H_2 0.2 l min^{-1}, Air 1 l min^{-1}, 20 A	152.49	147.18 (96.51)	155.32 (101.85)	8.14 (5.33)
	4,123	2,991 (72.53)	5,014 (121.60)	2,023 (49.06)
$\lambda_a = 1.1$, $\lambda_c = 1.3$, 15 A	158.36	154.28 (97.42)	162.00 (102.30)	7.72 (4.87)
	3,092	1,202 (38.68)	4,988 (160.55)	3,786 (121.86)
$\lambda_a = 1.3$, $\lambda_c = 2.5$, 15 A	154.08	149.99 (97.34)	157.61 (102.29)	7.62 (4.94)
	3,092	1,801 (58.04)	4,130 (133.09)	2,329 (75.05)
Simulation				
$\lambda_a = 1.3$, $\lambda_c = 2.5$, 10 A	160.00	158.57 (99.10)	162.13 (101.33)	3.56 (2.22)
	2,061	1,223 (49.58)	4,016 (162.78)	2,793 (113.20)
$\lambda_a = 1.3$, $\lambda_c = 2.5$, 15 A	160.00	158.36 (98.98)	162.78 (101.73)	4.41 (2.75)
	3,092	1,792 (54.92)	4,809 (147.40)	3,017 (92.47)
$\lambda_a = 1.3$, $\lambda_c = 2.5$, 20 A	160.00	158.33 (98.96)	164.31 (102.69)	5.97 (3.73)
	4,123	2,338 (55.82)	6,063 (144.81)	3,724 (88.96)
$\lambda_a = 1.3$, $\lambda_c = 2.5$, 25 A	160.00	158.30 (98.93)	166.06 (103.78)	7.76 (4.85)
	5,154	2,793 (51.02)	8,375 (152.99)	5,582 (101.97)
$\lambda_a = 1.3$, $\lambda_c = 2.5$, 50 A	160.00	158.53 (99.08)	172.26 (107.66)	13.73 (8.58)
	10,309	4,388 (43.51)	15,201 (150.68)	10,812 (107.2)

8.2.2 Type I flow-field – Operation with CO enriched hydrogen and air

Fig.42(a), a current of 15 A is drawn from the cell while operating with hydrogen and air. The distributions are the same as previously discussed. The overall current density distribution decreases almost linearly towards the cathode outlet, as the oxygen availability defines the shape of the distribution. The solid-phase temperature is higher in the region of the cathode inlet and is primarily defined by the shape of the current density (Fig.42(b)). When operating the cell at the same operating conditions with CO enriched hydrogen, the current density distribution changes. Fig.42(c) depicts the current density distribution for 0.5%CO in the hydrogen. The oxygen availability dictates the shape of the distribution, but a slight change is observed. The overall distribution appears to become flatter and increases in the regions of the cathode outlet and anode inlet and decreases in the regions of the cathode inlet and anode outlet. The solid-phase temperature distribution only changes slightly and is defined by the shape of the current density distribution (Fig.42(d)).

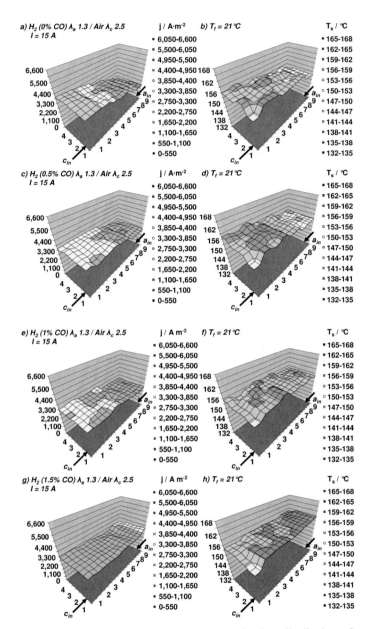

Fig.42. Measured solid-phase temperature and current density distribution. Set solid-phase temperature 160°C and gas inlet temperature 21°C. Stoichiometric flow rates for 15 A load current (0% CO (a,b), 0.5% CO (c,d), 1% CO (e,f), and 1.5% CO enriched hydrogen (g,h)).

For the case depicted in Fig.42(e), the oxygen availability still dictates the shape of the distribution, but the overall distribution becomes flatter. Again, the distribution increases in the region of the cathode outlet and anode inlet and decreases in the region of the cathode inlet and anode outlet. The solid-phase temperature distribution is defined by the shape of the current density distribution (Fig.42(f)). When operating the cell with 1.5%CO in the hydrogen, the current density distribution changes. The highest values are observed close to the anode inlet, and the lowest values occur within the middle region over the membrane electrode assembly area. The oxygen availability no longer determines the shape of the current density distribution. When increasing the CO content in the hydrogen, a current shift is observed from cathode inlet to anode inlet (Fig.42(g)). The same behaviour is observed for the solid-phase temperature distribution in Fig.42(h). The highest solid-phase temperature is observed in the region of the anode inlet, and the lowest temperature occurs in the region of the cathode inlet, whereas the mean temperature is higher than the previous value. Table 18 summarizes the measured results. The difference between the lowest and highest current densities decreased when CO enriched hydrogen was used at the anode side.

Table 18

Overview of measured solid-phase temperature and current density distribution for various operating conditions. Set solid-phase temperature and set current density equals 100%.

	Solid-phase temperature T_s and current density j			
	Mean	Minimum	Maximum	Difference
	°C	°C (%)	°C (%)	°C (%)
	A m^{-2}	A m^{-2} (%)	A m^{-2} (%)	A m^{-2} (%)
Measurement				
$\lambda_a = 1.3, \lambda_c = 2.5$, 15 A	154.08	149.99 (97.34)	157.61 (102.29)	7.62 (4.94)
	3,092	1,801 (58.04)	4,130 (133.09)	2,329 (75.05)
$\lambda_a = 1.3, \lambda_c = 2.5$, 15 A (0.5% CO)	154.63	151.39 (97.90)	158.23 (102.32)	6.84 (4.42)
	3,092	2,358 (76.11)	3,691 (119.15)	1,333 (43.03)
$\lambda_a = 1.3, \lambda_c = 2.5$, 15 A (1% CO)	155.24	152.73 (98.38)	157.46 (101.42)	4.73 (3.04)
	3,092	2,714 (87.29)	3,506 (112.77)	792 (25.47)
$\lambda_a = 1.3, \lambda_c = 2.5$, 15 A (1.5% CO)	156.16	151.73 (97.16)	158.67 (101.60)	6.94 (4.44)
	3,092	2,805 (90.00)	3,555 (114.07)	750 (24.06)

8.2.3 Type II flow-field – Operation with hydrogen and air

Fig.43 presents the results when operating the cell at stoichiometric flow rates of 1.3 (anode) and 5 (cathode). The load current is 15 A, and the set solid-phase temperature is 160°C. The highest current density is observed close to the cathode inlet, and the lowest value occurs in the middlemost gas channel region towards the cathode outlet. The current density on the right side over the membrane electrode assembly area is somewhat higher than that on the left side. The 5 outermost channels on both sides are in direct contact with both inlets and both outlets. The fluid-flow distribution within the gas channels and the resulting oxygen availability define the current density distribution. The solid-phase temperature is higher in the region of the cathode inlet and is primarily defined by the shape of the current density (Fig.43(b)). The air entering the cell at the cathode inlet influences the distribution only slightly. The highest solid-phase temperature is measured in the region of the cathode inlet. When drawing the same current from the cell and reducing the stoichiometric flow rate at the cathode side from 5 to 2.5, the current density distribution becomes more inhomogeneous (Fig.43(c)). The shape of the distribution follows the oxygen availability. Similar to the previous results, the values on the right side over the membrane electrode assembly area are somewhat higher than those on the left side. The highest solid-phase temperature is measured in the region of the cathode inlet, and the lowest solid-phase temperature towards the outlet, as shown in Fig.43(d). Fig.43(e) depicts the measurements obtained as the stoichiometric flow rate at the anode side is reduced from 1.3 to 1.1. The distribution becomes inhomogeneous and strictly follows the availability of oxygen and hydrogen. The highest overall values are seen on the right side over the membrane electrode assembly area, where both the oxygen and hydrogen are directly available due to the shape of the flow-field. In Fig.43(f), the lowest solid-phase temperature is located directly at the cathode outlet whereas higher overall values are seen in the region of the cathode inlet. Fig.43(g) returns the results obtained as the stoichiometric flow rate at the cathode side is reduced from 2.5 to 1.3. Higher values are observed on the right side over the membrane electrode assembly area, especially towards the cathode inlet. The lowest values are seen at the left side in the middle region over the membrane electrode assembly area. Similar to the previous measurements, the solid-phase temperature distribution follows the current density distribution, resulting in higher overall values close to the cathode inlet (Fig.43(h)). The lowest solid-phase temperature occurs in the region of the lowest current density and the highest value is located directly at the cathode inlet. Such stoichiometric flow rates should not be used when operating a HTPEM fuel cell, but in this case, they contribute to our understanding of the measured distributions.

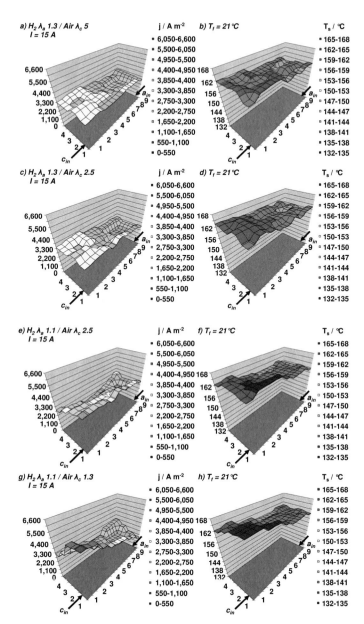

Fig.43. Measured solid-phase temperature and current density distribution. Set solid-phase temperature 160°C and gas inlet temperature 21°C. Stoichiometric flow rates for 15 A load current ($\lambda_a = 1.3$, $\lambda_c = 5$ (a,b), $\lambda_a = 1.3$, $\lambda_c = 2.5$ (c,d), $\lambda_a = 1.1$, $\lambda_c = 2.5$ (e,f), and $\lambda_a = 1.1$, $\lambda_c = 1.3$ (g,h)).

Selected simulation results are compared to the measurements. Fig.44(a) and Fig.44(b) present the current density and solid-phase temperature distribution within the reaction layer at reference operating conditions and 15A load current, respectively. The current density decreases from cathode inlet to cathode outlet as oxygen is continuously consumed due to ongoing electrochemical reactions. For this flow-field, higher oxygen mass fractions are observed in the region under the 5 outermost gas channels. The values are higher directly under the gas channels than under the land areas. A poor current density distribution is noticed in the middlemost gas channel region towards the cathode outlet. The distribution is strongly influenced by the bipolar-plate or flow-field structure and follows the fluid-flow distribution. Over the membrane electrode assembly area, higher current density values are observed under the ribs than compared to under the channels of the bipolar-plate. At the cathode inlet itself and under the lateral connector gas channel, the current density is slightly lower. A similar distribution was found for 20 A load current (Fig.44(c) and Fig.44(b)). The simulated solid-phase temperature agrees with the measurements. Within the porous media, the distribution is slightly influenced by the gases that enter the cell. The highest solid-phase temperature is seen in the region in which the highest current density is located. For 15 A load current, the total heat produced is 10.910 W within the reaction layer. Out of this value, the irreversible reaction heat comprises 70.79% (7.760 W), the reaction entropy comprises 28.81% (3.160 W), and the total Joule heating (resistive heating – electric and ionic) is 0.042 W. For 20 A load current, the total heat produced is 16.820 W (irreversible reaction heat 72.13% (12.190 W), the reaction entropy 27.30% (4.615 W), resistive heating 0.55% (0.094 W)). At 25 A load current, the total heat produced increased to 23.950 W are produced (irreversible reaction heat 73.15%, the reaction entropy 26.10%, resistive heating 0.74%).

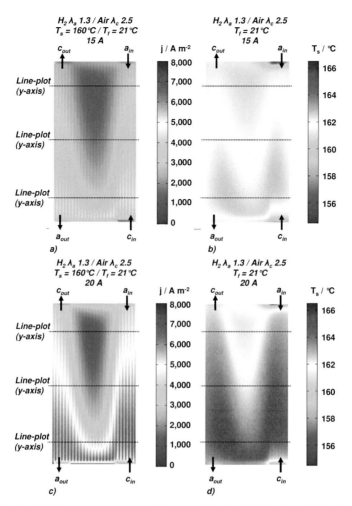

Fig.44. Simulated solid-phase temperature and current density distribution within the cathode side reaction layer (slice plot in x-y-plane, $z = -379$ μm). Set solid-phase temperature 160°C and gas inlet temperature 21°C. Stoichiometric flow rates for 15 A (a,b) and 20 A load current (c,d).

Fig.45 depicts the current density for different load currents as a line-plot. Close to the cathode inlet, the current density under the land is considerably higher than that under the channel. The observed difference becomes less pronounced towards the cathode outlet for all load currents. As the load current increases, the difference between the lowest and highest current density also increases. As mentioned above, the low current density values in the middlemost gas channel region are directly related to the bad fluid-flow distribution and oxygen availability, as shown in Fig.19. The simulation results are compared to the segmented measurements for 15 A load current. For the three line-plots along the y-axis (close to the cathode inlet, in the middle over the membrane electrode assembly area, and towards the cathode outlet), both distributions show a similar shape. The measured current density gradient is not as pronounced as the simulated ones.

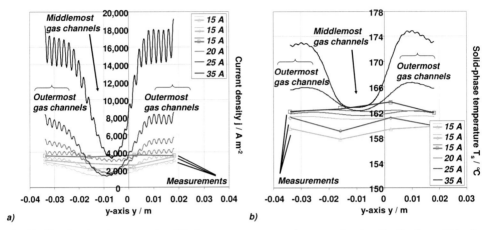

Fig.45. Simulated and measured solid-phase temperature and current density distribution within the cathode side reaction layer (three line-plots along the y-axis as indicated in Fig.44 ($z = -379$ µm)). Set solid-phase temperature 160°C and gas inlet temperature 21°C. Stoichiometric flow rates for 15 A, 20 A, 25A, and 35 A load current.

For this flow-field, the lower current density values close to the cathode inlet may be explained as follows. The water vapor mass fraction increases towards the cathode outlet according to the shape of the flow-field and towards the reaction layer (z-axis). Higher values are observed under the land areas, whereas the amount increases with the load current. Similar to type I flow-field, the mean cathode side H_3PO_4 concentration varies, ranging from 96.13wt.% (1 A load current), to 97.50wt.% (10 A load current) and 96.04wt.% (at 35 A load current). For the distribution over the membrane electrode assembly area, higher concentration values are found close to the cathode inlet and the outermost channels, whereas lower values occur in the middlemost gas channel region towards the cathode outlet. At the anode side the phosphoric acid concentration decreases almost linearly from 107.80wt.% (1 A load current) to 104.46wt.% (35 A load current). Similar as for type I flow-field, the oxygen diffusion coefficient is lowest in the region close to the inlet, whereas it remains approximately constant elsewhere. The oxygen solubility is approximately constant for all load currents. The simulated solid-phase temperature distribution in Fig.45 follows the current density distribution and fluid-flow distribution. Within the porous media, the distribution is influenced by the gases entering the cell with a fluid-(gas)-phase temperature of 21°C. This trend is observed at both gas inlets and in the region under the lateral gas channels and is slightly more pronounced at the cathode side. The highest solid-phase temperature is observed close to the region in which the highest current density is located, and the lowest temperature occurs in the middlemost gas channel region towards the cathode outlet. Due to the higher current density under the land area, the solid-phase temperature peaks accordingly. Additionally, the temperature in the regions under the gas channels is somewhat lower than that under the land. This temperature difference almost vanishes towards the exit. The same trend is observed for the current density distribution, and the solid-phase temperature for 15 A load current is compared to the simulations at three different positions over the membrane electrode assembly area. Overall, the reported values agree well with the simulations. At 15 A load current, the highest solid-phase temperature is 162.83°C, at 20 A it is 164.44°C, at 25 A it is 166.85°C, and at 35 A it is 174.51°C. These values are somewhat higher than the reported values for type I flow-field.

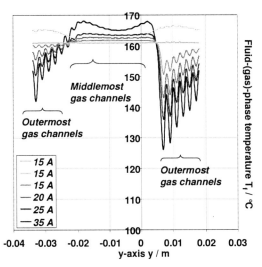

Fig.46. Simulated fluid-(gas)-phase temperature distribution within the cathode side gas diffusion layer. Set solid-phase temperature 160°C and gas inlet temperature 21°C (three line-plots along the y-axis as indicated in Fig.44 ($z = -1$ μm)). Stoichiometric flow rates for 15 A, 20 A, 25A, and 35 A load current.

Simulated fluid-(gas)-phase temperatures for different load currents are shown in Fig.46 as line-plots. The temperature distribution increases towards the cathode outlet and becomes considerably flatter as it reaches thermal equilibrium. For all load currents, the fluid-(gas)-phase temperature is low in the region of the outermost channels and high in the middlemost gas channel region. As more current is drawn, more gas enters the cell due to the stoichiometric flow rates, and the gas takes longer to reach the set solid-phase temperature. Thermal equilibrium is quickly achieved at the anode side.

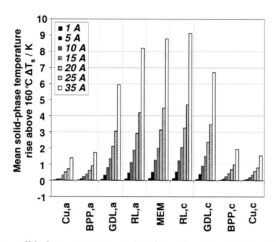

Fig.47. Simulated mean solid-phase temperature rise above the set solid-phase temperature within the different components. Set solid-phase temperature 160°C and gas inlet temperature 21°C. Stoichiometric flow rates for 1 A, 5 A, 10 A, 15 A, 20 A, 25A, and 35 A load current.

Fig.47 depicts the simulated mean solid-phase temperature rise above 160°C for the anode and cathode side components up to 35 A load current. Table 19 summarizes the simulated heat transfer between the two phases for selected load currents. Within the gas channels, the heat is absorbed by the gas and transported out of the cell. Similar to type I flow-field, more heat is transferred between the two phases at the cathode side.

Table 19

Simulated heat transfer between the two phases / W

	Load current I / A			
	15	20	25	35
Anode side component				
Gas channel	0.258	0.34	0.428	0.595
GDL	0.2736	0.3495	0.4301	0.621
RL	0.09	0.1219	0.1765	0.3136
Cathode side component				
Gas channel	1.421	1.933	2.45	3.551
GDL	0.3717	0.502	0.642	1.372
RL	0.0541	0.0755	0.0993	0.161

Table 20 summarizes the simulated and measured results. As the load current is increased, the difference also increases, leading to more inhomogeneous quantities distributions.

Table 20

Overview of simulated and measured solid-phase temperature and current density distribution for various operating conditions. Set solid-phase temperature and set current density equals 100%.

	Solid-phase temperature T_s and current density j			
	Mean	Minimum	Maximum	Difference
	°C	°C (%)	°C (%)	°C (%)
	A m^{-2}	A m^{-2} (%)	A m^{-2} (%)	A m^{-2} (%)
Measurement				
$\lambda_a = 1.3$, $\lambda_c = 5$, 15 A	159.92	153.19 (95.79)	163.44 (102.20)	10.25 (6.40)
	3,092	2,248 (72.10)	3,762 (120.66)	1,514 (48.55)
$\lambda_a = 1.3$, $\lambda_c = 2.5$, 15 A	161.09	156.02 (96.85)	164.87 (102.35)	8.85 (5.49)
	3,092	1,879 (61.68)	3,898 (127.96)	2,019 (66.27)
$\lambda_a = 1.1$, $\lambda_c = 2.5$, 15 A	161.19	158.02 (98.03)	165.74 (102.82)	7.72 (4.78)
	3,092	1,754 (56.68)	4,410 (142.53)	2,656 (85.84)
$\lambda_a = 1.1$, $\lambda_c = 1.3$, 15 A	161.67	158.28 (97.90)	167.71 (103.73)	9.43 (5.83)
	3,092	1,797 (58.03)	5,537 (178.84)	3,740 (120.80)
$\lambda_a = 1.3$, $\lambda_c = 2.5$, 20 A	159.03	152.86 (96.11)	162.36 (102.09)	9.5 (5.97)
	4,123	2,907 (70.40)	5,179 (125.42)	2,272 (55.02)
Simulation				
$\lambda_a = 1.3$, $\lambda_c = 2.5$, 10 A	160.00	160.8 (100.51)	161.67 (101.04)	0.85 (0.53)
	2,061	1,133 (59.93)	2,729 (144.33)	1,595 (84.39)
$\lambda_a = 1.3$, $\lambda_c = 2.5$, 15 A	160.00	161.2 (100.76)	162.83 (101.77)	1.61 (1.01)
	3,092	1,274 (44.85)	4,431 (155.99)	3,157 (111.14)
$\lambda_a = 1.3$, $\lambda_c = 2.5$, 20 A	160.00	161.5 (100.98)	164.44 (102.77)	2.85 (1.78)
	4,123	1,466 (35.34)	6,642 (160.12)	5,176 (124.78)
$\lambda_a = 1.3$, $\lambda_c = 2.5$, 25 A	160.00	161.8 (101.15)	166.85 (104.28)	5.01 (3.12)
	5,154	1,295 (23.25)	9,980 (179.24)	8,685 (155.98)
$\lambda_a = 1.3$, $\lambda_c = 2.5$, 35 A	160.00	163.2 (102.00)	174.51 (109.07)	11.31 (7.06)
	7,216	3,301 (38.35)	19,166 (222.65)	15,864 (184.3)

8.2.4 Type II flow-field – Operation with CO enriched hydrogen and air

In Fig.48(a), a load current of 20 A is drawn from the cell while operating with hydrogen and air. The distributions are highly similar to those shown in Fig.43(c) and Fig.43(d). Fig.48(c) depicts the current density distribution for 20 A load current and 2%CO in the hydrogen. The shape of the current density distribution is defined by the oxygen availability and is overlapped by the hydrogen availability. The current density decreases in the region of the anode outlet and increases in the region of the anode inlet. The highest overall values are observed on the right side over the membrane electrode assembly area towards the cathode inlet, whereas the lowest values occur in the region of the anode outlet and in the middlemost channel region towards the cathode outlet. The solid-phase temperature distribution follows the current density distribution and is similar to previous results. Fig.48 also displays the distributions measured as the CO content in the hydrogen is increased to 4% and 6%. In Fig.48(e) the oxygen availability no longer dictates the shape of the distribution. The highest overall values are observed on the right side over the membrane electrode assembly area. The current density increases in the region of the anode inlet and decreases in the region of the anode outlet. The solid-phase temperature distribution changes in accordance with the current density distribution. As the CO content is increased to 6%, the hydrogen availability completely dictates the shape of the current density distribution (Fig.48(g)). The highest values are observed close to the anode inlet, whereas the lowest values arise close to the anode outlet. Similar to the previous measurements for type I flow-field, when increasing the CO content in the hydrogen, a current shift is observed towards the anode inlet. The same behaviour is observed for the solid-phase temperature distribution shown in Fig.48(h). The highest solid-phase temperature occurs in the region of the anode inlet, and the lowest is observed in the region of the cathode inlet. The mean temperature is higher than the previous results.

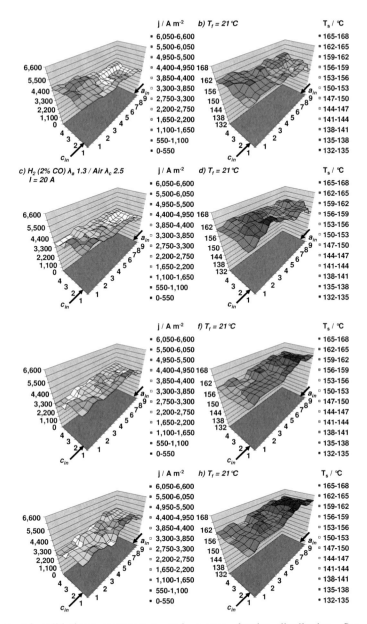

Fig.48. Measured solid-phase temperature and current density distribution. Set solid-phase temperature 160°C and gas inlet temperature 21°C. Stoichiometric flow rates for 20 A load current (0% CO (a,b), 2% CO (c,d), 4% CO (e,f), and 6% CO enriched hydrogen (g,h)).

Table 21 summarizes the measured solid-phase temperature and current density distribution for selected operating conditions. Similar to the results presented for type I flow-field, the difference between the lowest and highest current density decreases when CO enriched hydrogen is used at the anode side.

Table 21

Overview of measured solid-phase temperature and current density distribution for various operating conditions. Set solid-phase temperature and set current density equals 100%.

	Solid-phase temperature T_s and current density j			
	Mean	Minimum	Maximum	Difference
	°C	°C (%)	°C (%)	°C (%)
	A m^{-2}	A m^{-2} (%)	A m^{-2} (%)	A m^{-2} (%)
Measurement				
$\lambda_a = 1.3$, $\lambda_c = 2.5$, 20 A	159.03	152.86 (96.11)	162.36 (102.09)	9.5 (5.97)
	4,123	2,907 (70.40)	5,179 (125.42)	2,272 (55.02)
$\lambda_a = 1.3$, $\lambda_c = 2.5$, 20 A (2% CO)	160.67	155.87 (97.01)	166.55 (103.66)	10.68 (6.64)
	4,123	3,234 (78.35)	5,115 (123.93)	1,881 (45.57)
$\lambda_a = 1.3$, $\lambda_c = 2.5$, 20 A (4% CO)	161.25	157.01 (97.36)	166.51 (103.26)	9.50 (5.89)
	4,123	3,238 (78.50)	5,004 (121.32)	1,766 (42.81)
$\lambda_a = 1.3$, $\lambda_c = 2.5$, 20 A (6% CO)	162.48	158.19 (97.35)	168.00 (103.39)	9.81 (6.03)
	4,123	3,043 (73.46)	5,284 (127.56)	2,241 (54.10)

8.2.5 Type III flow-field – Operation with hydrogen and air

Fig.49(a) and Fig.49(b) display the results for the cell with type III flow-field operating with hydrogen and air at stoichiometric flow rates of 1.3 (anode) and 2.5 (cathode) when drawing 15 A load current. The highest current density is observed close to the cathode inlet, and the lowest value occurs close to the cathode outlet. The current density on the right side over the membrane electrode assembly area is higher than that on the left side over the membrane electrode assembly area. The right side is in direct contact with the cathode inlet. The oxygen availability primarily defines the current density distribution. The solid-phase temperature shown in Fig.49(b) follows the current density distribution and exhibits high values close to the cathode inlet and lower values in the region towards the cathode outlet. Fig.49(c) and Fig.49(d) depict the results obtained as the stoichiometric flow rate at the anode side is reduced from 1.3 to 1.1 while all other parameters are held constant. The influence on the quantities distribution is weak in this case. Thus, the current density distribution is very stable against changes in the anode stoichiometric flow rates.

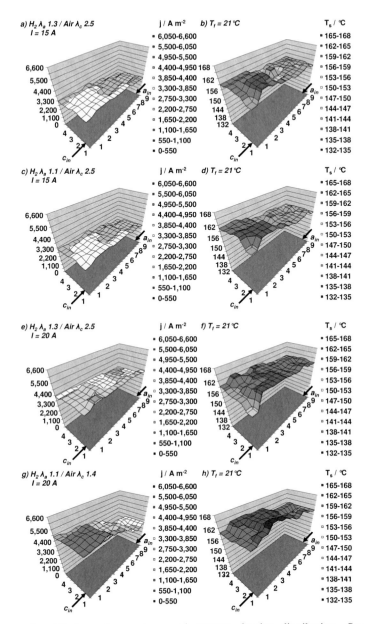

Fig.49. Measured solid-phase temperature and current density distribution. Set solid-phase temperature 160°C and gas inlet temperature 21°C. Stoichiometric flow rates for 15 A ($\lambda_a = 1.3$, $\lambda_c = 2.5$ (a,b), $\lambda_a = 1.1$, $\lambda_c = 2.5$ (c,d)) and 20 A load current ($\lambda_a = 1.3$, $\lambda_c = 2.5$ (e,f), $\lambda_a = 1.1$, $\lambda_c = 1.4$ (g,h)).

The current density distribution and solid-phase temperature displayed in Fig.49(e) and Fig.49(f) follow a distribution similar to those of Fig.49(a) and Fig.49(b). The highest solid-phase temperature is located in the region of the highest current density. In Fig.49(g), the stoichiometric flow rates are reduced to 1.1 (anode) and 1.4 (cathode). The oxygen availability dictates the shape of the current density distribution. A large gradient is observed between the cathode inlet and the cathode outlet. The solid-phase temperature distribution in Fig.49(h) does not significantly change from the previous measurements but the mean value increases. The influence of the air entering the cell vanishes for low stoichiometric flow rates.

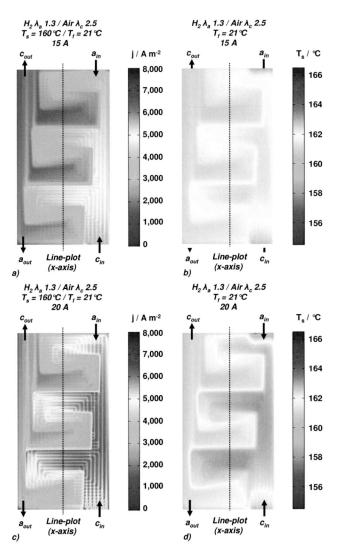

Fig.50. Simulated solid-phase temperature and current density distribution within the cathode side reaction layer (slice plot in x-y-plane, $z = -379$ μm). Set solid-phase temperature 160°C and gas inlet temperature 21°C. Stoichiometric flow rates for 15 A (a,b) and 20 A load current (c,d).

Fig.50 presents the simulated solid-phase temperature and current density distribution at given operating conditions when drawing 15 A and 20 A load currents. In accordance with the measurements (Fig.49(a) and Fig.49(b)), the shape of the current density is dominated by the oxygen availability. High current density values are observed on the right side over the membrane electrode assembly area under the gas channels that are in direct contact with the cathode inlet. The current density decreases towards the cathode outlet. The solid-phase temperature follows the current density distribution, and higher values are found on the right side. Fig.50(c) and Fig.50(d) present the simulated results obtained when drawing 20 A load current. The results are similar to those for 15 A load current. The shape of the current density is dominated by the oxygen availability. High current density values are observed on the right side over the membrane electrode assembly area under the gas channels that are in direct contact with the cathode inlet. The current density decreases towards the cathode outlet. The solid-phase temperature follows the current density distribution and high values are found on the right side and partially in the middlemost region over the membrane electrode assembly area. For 15 A load current, the total heat produced within the cathode side reaction layer is 11.43 W. Of this value, the irreversible reaction heat comprises 68.44% (7.86 W), the reaction entropy comprises 31.07% (3.56 W), and the total Joule heating (resistive heating – electric and ionic) makes up the remaining 0.48% (0.055 W).

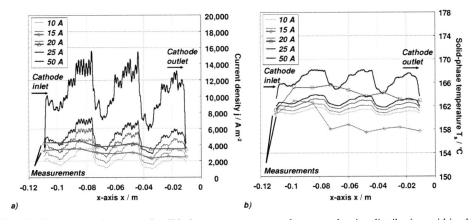

Fig.51. Simulated and measured solid-phase temperature and current density distribution within the cathode side reaction layer (line-plot along the x-axis as indicated in Fig.50 ($z = -379$ μm)). Set solid-phase temperature 160°C and gas inlet temperature 21°C. Stoichiometric flow rates for 10 A, 15 A, 20 A, 25A, and 50 A load current.

The same trend is observed for 20 A load current where 15.680 W are produced (irreversible reaction heat 69.25% (10.91 W), the reaction entropy 30.15% (4.75 W), resistive heating 0.58% (0.092 W)) and 25 A load current where 19.97 W are produced (irreversible reaction heat 69.91% (14.03 W), the reaction entropy 29.38% (5.89 W), resistive heating 0.69% (0.139 W)). The reported values for type III flow-field are in agreement with the values found for type I and type II flow-fields. Fig.51 presents the simulated solid-phase temperature and current density distribution for 10 A, 15 A, 20 A, 25 A, and 50 A load currents as a line-plot. The oxygen mass fraction behaves exactly as explained for the other two types of flow-fields. The difference between the current density values under the land and channel areas increase for high load currents. Moreover, the current density decreases slightly towards the cathode outlet. The three regions of the membrane electrode assembly area can be clearly distinguished. The simulation results for 15 A and 20 A load currents are compared to the measurements. Similar to the other two types of flow-fields, the gradient in the measurements is not as large as the simulated result. The low current density values close to the cathode inlet can be explained as follows. The water vapor mass fraction displays the same overall behaviour as discussed for the other two types of flow-fields. The fraction increases towards the cathode outlet according to the shape of the flow-field and towards the reaction layer (z-axis). High values are observed under the land areas and increase with increasing load current. The mean cathode side H_3PO_4 concentration is minimal at 1 A load current (94.92wt.%). The concentration increases to 96.45wt.% (25 A load current) and decreases to 95.81wt.% at 50 A load current. The distribution over the membrane electrode assembly area, high concentration values were found in the middle of the flow-field. At the anode side, the phosphoric acid concentration decreases almost linearly from 106.44wt.% (1 A load current) to 101.20wt.% (50 A load current). Again similar to type I and type II flow-field, the oxygen diffusion coefficient is lowest in the region close to the inlet whereas it remains approximately constant elsewhere. The oxygen solubility decreases for 1 A load current to 25 A load current and increases slightly for 50 A load current. Thus, the solubility changes according to the phosphoric acid concentration at the cathode side. The simulated solid-phase temperature distribution for various load currents follows the current density distribution and exhibits a decreasing trend from cathode inlet to cathode outlet (Fig.51(b)). Additionally, the three regions of the flow-field can clearly be distinguished. The simulation results are compared to the measurements for 15 A and 20 A load currents. The segmentation is not sufficiently fine to account for the simulated gradient, but the fluctuations between the three regions of the flow-field can be observed.

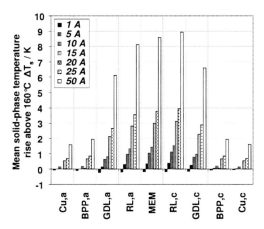

Fig.52. Simulated mean solid-phase temperature rise above the set solid-phase temperature within the different components. Set solid-phase temperature 160°C and gas inlet temperature 21°C. Stoichiometric flow rates for 1 A, 5 A, 10 A, 15 A, 20 A, 25A, and 50 A load current.

Fig.52 depicts the simulated mean solid-phase temperature rise above 160°C for all components for load currents ranging from 1 A to 50 A. The solid-phase temperature increase in the gold-plated copper current collectors and the bipolar-plates is quite low. The mean temperature rise is more pronounced within the gas diffusion layer, the reaction layer, and the membrane. For a load current of 50 A, the highest solid-phase temperature rise is 8.94°C, occurring within the cathode side reaction layer.

Table 22

Simulated heat transfer between the two phases / W

	Current density I / A			
	15	20	25	50
Anode side component				
Gas channel	0.275	0.395	0.517	1.174
GDL	0.258	0.34	0.421	0.866
RL	0.0539	0.0737	0.0928	0.199
Cathode side component				
Gas channel	1.33	1.774	2.21	4.486
GDL	0.383	0.519	0.655	1.371
RL	0.0532	0.073	0.0922	0.196

For a load current of 1 A, the mean solid-phase temperature decreases slightly because more heat is transferred out of the cell than is produced within the cell. The values reported for type III flow-field are in agreement with those for type I flow-field but are lower than those for type II flow-field, as shown in Fig.47. The simulated fluid-(gas)-phase temperature is not shown here. Its behaviour is similar to that discussed for type I and type II flow-field. Air enters the HTPEM fuel cell with 21°C and is heated up until thermal equilibrium is reached. Depending on the load current, it takes longer for equilibrium to be achieved because more heat has to be exchanged between the two phases. Table 22 summarizes the simulated heat transfer between the two phases for selected load currents. Within the gas channels, the heat is absorbed by the gas and transported out of the cell. Similar to type I and type II flow-fields, more heat is transferred between the two phases at the cathode side.

Table 23

Overview of simulated and measured solid-phase temperature and current density distribution for various operating conditions. Set solid-phase temperature and set current density equals 100%.

	Solid-phase temperature T_s and current density j			
	Mean	Minimum	Maximum	Difference
	°C	°C (%)	°C (%)	°C (%)
	A m^{-2}	A m^{-2} (%)	A m^{-2} (%)	A m^{-2} (%)
Measurement				
$\lambda_a = 1.1, \lambda_c = 2.5$, 15 A	158.45	155.11 (97.89)	162.99 (102.86)	7.88 (4.97)
	3,092	2,236 (72.35)	3,745 (121.18)	1,509 (48.82)
$\lambda_a = 1.3, \lambda_c = 2.5$, 15 A	157.93	153.20 (97.00)	163.10 (103.27)	9.90 (6.26)
	3,092	2,237 (71.94)	3,785 (121.72)	1,548 (49.78)
$\lambda_a = 1.1, \lambda_c = 1.4$, 20 A	161.25	154.60 (95.87)	166.87 (103.48)	12.27 (7.60)
	4,123	2,410 (58.48)	5,333 (129.43)	2,923 (70.94)
$\lambda_a = 1.3, \lambda_c = 2.5$, 20 A	160.49	155.21 (96.70)	163.08 (101.61)	7.87 (4.90)
	4,123	3,374 (81.72)	4,906 (118.83)	1,532 (37.10)
Simulation				
$\lambda_a = 1.3, \lambda_c = 2.5$, 10 A	160.00	158.57 (99.10)	162.13 (101.33)	3.56 (2.22)
	2,061	1,053 (42.68)	3,628 (147.04)	2,574 (104.35)
$\lambda_a = 1.3, \lambda_c = 2.5$, 15 A	160.00	158.36 (98.98)	162.78 (101.73)	4.41 (2.75)
	3,092	1,556 (47.69)	4,771 (146.21)	3,214 (98.52)
$\lambda_a = 1.3, \lambda_c = 2.5$, 20 A	160.00	158.33 (98.96)	164.31 (102.69)	5.97 (3.73)
	4,123	2,185 (52.19)	6,085 (145.34)	3,900 (93.15)
$\lambda_a = 1.3, \lambda_c = 2.5$, 25 A	160.00	158.30 (98.93)	166.06 (103.78)	7.76 (4.58)
	5,154	2,898 (52.94)	7,398 (135.15)	4,500 (82.21)
$\lambda_a = 1.3, \lambda_c = 2.5$, 50 A	160.00	158.53 (99.08)	172.26 (107.66)	13.73 (8.58)
	10,309	3,687 (36.55)	15,560 (154.24)	11,873 (117.6)

Table 23 summarizes the simulated and measured results. Overall, the minimal difference for the solid-phase temperature and the current density distribution occurs for 15 A and 20 A load current. The difference increases as the load current is increased, leading to more inhomogeneous quantities distributions.

8.2.6 Type III flow-field – Operation with CO enriched hydrogen and air

Table 24 summarizes the measured solid-phase temperature and current density distributions. The current density distribution becomes more homogeneous for a cell operating with CO enriched hydrogen. The corresponding distributions are shown in Fig.66 and Fig.68 as part of the segmented EIS measurements.

Table 24

Overview of measured solid-phase temperature and current density distribution for various operating conditions. Set solid-phase temperature and set current density equals 100%.

	Solid-phase temperature T_s and current density j			
	Mean	Minimum	Maximum	Difference
	°C	°C (%)	°C (%)	°C (%)
	A m^{-2}	A m^{-2} (%)	A m^{-2} (%)	A m^{-2} (%)
Measurement				
$\lambda_a = 1.3, \lambda_c = 2$, 20 A	160.32	156.50 (97.61)	162.33 (101.25)	5.83 (3.63)
	4,123	3,044 (73.45)	5,106 (123.21)	2,062 (49.75)
$\lambda_a = 1.3, \lambda_c = 2.5$, 20 A	160.49	155.21 (96.70)	163.08 (101.61)	7.87 (4.90)
	4,123	3,374 (81.72)	4,906 (118.83)	1,532 (37.10)
$\lambda_a = 1.3, \lambda_c = 2.5$, 20 A (2% CO)	161.25	153.86 (95.41)	163.79 (101.57)	9.93 (6.15)
	4,123	3,681 (89.10)	4,731 (114.51)	1,050 (25.41)
$\lambda_a = 1.3, \lambda_c = 2.5$, 20 A (4% CO)	162.43	157.15 (96.74)	166.86 (102.73)	9.71 (5.97)
	4,123	3,571 (86.46)	4,634 (112.21)	1,063 (25,74)
$\lambda_a = 1.3, \lambda_c = 2.5$, 20 A (6% CO)	163.86	159.52 (97.35)	167.23 (102.06)	7.71 (4.70)
	4,123	3,285 (79.60)	4,900 (118.74)	1,615 (39.13)

8.3 Load operating conditions – Co-flow configuration

8.3.1 Type I flow-field – Operation with hydrogen and air

Fig.53(a) presents results for the cell operating with hydrogen and air at stoichiometric flow rates of 1.3 (anode) and 2.5 (cathode) when drawing 15 A load current. The highest current density is observed close to the cathode inlet, and the lowest is observed at the cathode outlet. The overall current density distribution decreases almost linearly from cathode inlet to cathode outlet. The current density gradient is larger than the gradient shown in Fig.42 for the same operating conditions in counter-flow configuration. The solid-phase temperature is high in the region of the cathode inlet and is defined by the shape of the current density (Fig.53(b)). The highest temperature is measured in the region of the cathode inlet, whereas the lowest temperature is measured towards the cathode outlet. When decreasing the stoichiometric flow rate at the cathode side from 2.5 to 1.3, the current density becomes more inhomogeneous (Fig.53(c)). The oxygen availability defines the current density shape and the highest values are measured at the cathode inlet and the lowest values are observed at the cathode outlet. Again, the highest temperature is measured in the region of the cathode inlet (Fig.53(d)).

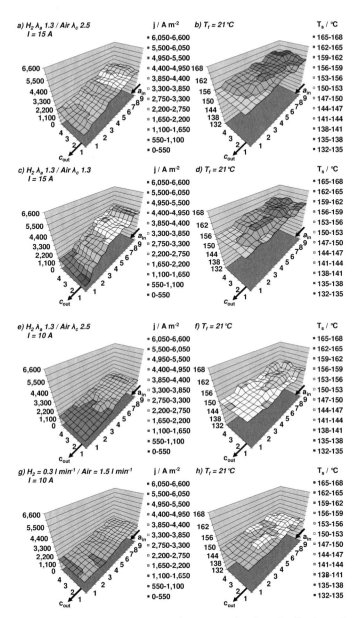

Fig.53. Measured solid-phase temperature and current density distribution. Set solid-phase temperature 160°C and gas inlet temperature 21°C. Stoichiometric flow rates for 15 A (λ_a = 1.3, λ_c = 2.5 (a,b), λ_a = 1.3, λ_c = 1.3 (c,d)), 10 A (λ_a = 1.3, λ_c = 2.5 (e,f)), and constant flow rates for 10 A load current (g,h).

In Fig.53(e), 10 A load current is drawn from the cell at stoichiometric flow rates of 1.3 (anode) and 2.5 (cathode). The almost linear decrease of the current density is similar to the trend observed for 15

A load current, as previously discussed. The solid-phase temperature distribution follows the current density distribution (Fig.53(f)). As the flow rates are increased while keeping the load current at 10 A, the current density distribution becomes flatter (Fig.53(g)). As for the solid-phase temperature, the mean value decreases as can be seen in Fig.53(h). The lowest solid-phase temperature occurs at the cathode inlet, whereas the highest solid-phase temperature is observed in the region of the highest current density. The solid-phase temperature gradient is relatively small over the membrane electrode assembly area due to the high flow rates and the shape of the flow-field. Table 25 summarizes the results for the above measurements.

Table 25

Overview of measured solid-phase temperature and current density distribution for various operating conditions. Set solid-phase temperature and set current density equals 100%.

	Solid-phase temperature T_s and current density j			
	Mean	Minimum	Maximum	Difference
	°C	°C (%)	°C (%)	°C (%)
	A m^{-2}	A m^{-2} (%)	A m^{-2} (%)	A m^{-2} (%)
Measurement				
H$_2$ 0.3 l min^{-1}, Air 1.5 l min^{-1}, 10 A	149.65	145.90 (97.48)	152.32 (101.77)	6.42 (4.28)
	2,061	1,581 (76.65)	2,610 (126.53)	1,029 (49.88)
$\lambda_a = 1.3$, $\lambda_c = 2.5$, 10 A	152.50	149.64 (98.11)	156.25 (102.45)	6.61 (4.33)
	2,061	1,149 (55.75)	2,825 (137.09)	1,676 (81.33)
$\lambda_a = 1.3$, $\lambda_c = 1.3$, 15 A	158.64	154.55 (97.41)	162.64 (102.53)	8.12 (5.11)
	3,092	862 (27.68)	4,658 (149.61)	3,796 (121.92)
$\lambda_a = 1.3$, $\lambda_c = 2.5$, 15 A	158.26	155.05 (97.97)	161.48 (102.03)	6.43 (4.06)
	3,092	1,767 (56.99)	4,211 (135.82)	2,444 (78.82)

8.3.2 Type I flow-field – Operation with CO enriched hydrogen and air

Fig.54 presents the current density distribution for 20 A load current which has been previously discussed. When increasing the CO content in the hydrogen to 0.5%, 1%, and 1.5% the same current density shift from anode outlet to anode the inlet is observed, as previously discussed.

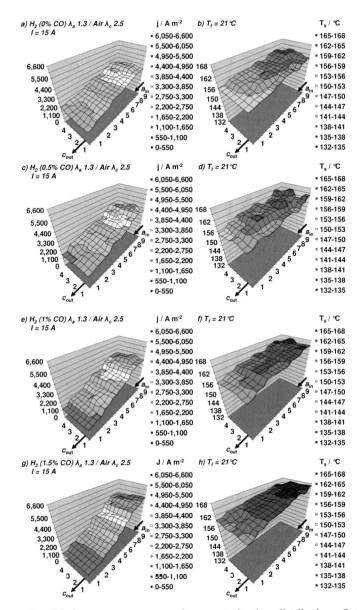

Fig.54. Measured solid-phase temperature and current density distribution. Set solid-phase temperature 160°C and gas inlet temperature 21°C. Stoichiometric flow rates for 15 A load current (0% CO (a,b), 0.5% CO (c,d), 1% CO (e,f), and 1.5% CO enriched hydrogen (g,h)).

The oxygen and hydrogen availabilities define the shape of the distribution. The difference between the highest and lowest values is considerably larger than that observed for the same operating conditions in counter-flow configuration (Fig.42). The solid-phase temperature distribution follows the current density distribution. Table 26 summarizes the measured results. The difference is generally higher for co-flow configuration than for counter-flow configuration. Additionally, in contrast to the results for counter-flow configuration, the difference between the lowest and highest current density value increases when CO enriched hydrogen is used at the anode side.

Table 26

Overview of measured solid-phase temperature and current density distribution for various operating conditions. Set solid-phase temperature and set current density equals 100%.

	Solid-phase temperature T_s and current density j			
	Mean	Minimum	Maximum	Difference
	°C	°C (%)	°C (%)	°C (%)
	A m^{-2}	A m^{-2} (%)	A m^{-2} (%)	A m^{-2} (%)
Measurement				
$\lambda_a = 1.3$, $\lambda_c = 2.5$, 15 A	158.26	155.05 (97.97)	161.48 (102.03)	6.43 (4.06)
	3,092	1,767 (56.99)	4,211 (135.82)	2,444 (78.82)
$\lambda_a = 1.3$, $\lambda_c = 2.5$, 15 A (0.5% CO)	159.94	154.82 (96.79)	164.85 (103.06)	10.03 (6.27)
	3,092	1,544 (49.90)	4,596 (148.54)	3,052 (98.64)
$\lambda_a = 1.3$, $\lambda_c = 2.5$, 15 A (1% CO)	161.82	156.82 (96.90)	166.81 (103.07)	9.99 (6.17)
	3,092	1,294 (41.61)	4,758 (153.01)	3,464 (111.40)
$\lambda_a = 1.3$, $\lambda_c = 2.5$, 15 A (1.5% CO)	163.25	158.09 (96.83)	166.88 (102.21)	8.79 (5.38)
	3,092	1,164 (37.62)	5,175 (167.28)	4,011 (129.65)

8.3.3 Type II flow-field – Operation with hydrogen and air

Fig.55(a) and Fig.55(b) present results for the cell with type II flow-field operating with hydrogen and air at stoichiometric flow rates of 2.6 (anode) and 5 (cathode). The current density on the left side over the membrane electrode assembly area is higher than that on the right side because the 5 outermost channels are in direct contact with both inlets and both outlets. The fluid-flow distribution within the gas channels and the resulting oxygen availability define the current density distribution. When drawing 15 A load current from the cell and reducing the stoichiometric flow rates, the current density distribution becomes more inhomogeneous (Fig.55(c)). The solid-phase temperature distribution exhibits a typical shape, as previously discussed (Fig.55(d)).

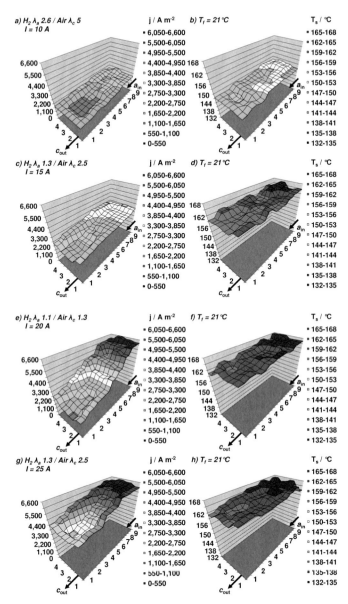

Fig.55. Measured solid-phase temperature and current density distribution. Set solid-phase temperature 150°C and 160°C and gas inlet temperature 21°C. Stoichiometric flow rates for 10 A (λ_a = 2.6, λ_c = 5 (a,b)), 15 A (λ_a = 1.3, λ_c = 2.5 (c,d)), 20 A (λ_a = 1.1, λ_c = 1.3 (e,f)), and 25 A load current (λ_a = 1.3, λ_c = 2.5 (g,h)).

Fig.55(e) depicts the current density distribution for 20A load current. The distribution follows the availability of oxygen and hydrogen and is highly inhomogeneous. The solid-phase temperature distribution in Fig.55(f) follows the current density distribution. Fig.55(g) displays the results for a cell operating at stoichiometric flow rates of 1.3 (anode) and 2.5 (cathode) while drawing 25 A from the cell. The shape of the current density distribution is similar to the distribution at other load currents for the same stoichiometric flow rates. The solid-phase temperature distribution follows the current density distribution, resulting in high overall values close to the cathode inlet (Fig.55(h)). Table 27 summarizes the measured results. The difference is generally higher for co-flow configuration than for counter-flow configuration. Table 27 also presents the results when operating the cell with CO enriched hydrogen. The corresponding measured distributions are depicted in Fig.60. In contrast to the results for counter-flow configuration, the difference between the lowest and highest current density values increases when CO enriched hydrogen is used at the anode side.

Table 27

Overview of measured solid-phase temperature and current density distribution for various operating conditions. Set solid-phase temperature and set current density equals 100%.

	Solid-phase temperature T_s and current density j			
	Mean	Minimum	Maximum	Difference
	°C	°C (%)	°C (%)	°C (%)
	A m^{-2}	A m^{-2} (%)	A m^{-2} (%)	A m^{-2} (%)
Measurement				
$\lambda_a = 2.6$, $\lambda_c = 5$, 10 A	149.43	145.76 (97.54)	152.50 (102.05)	6.74 (4.51)
	2,061	1,424 (68.19)	2,660 (127.37)	1,236 (59.18)
$\lambda_a = 1.1$, $\lambda_c = 1.3$, 20 A	164.31	160.49 (97.67)	167.97 (102.22)	7.48 (4.55)
	4,123	1,584 (38.43)	6,596 (160.05)	5,012 (121.61)
$\lambda_a = 1.3$, $\lambda_c = 2.5$, 25 A	164.18	159.96 (97.42)	167.55 (102.04)	7.59 (4.62)
	5,154	3,233 (62.61)	6,470 (125.29)	3,237 (62.68)
$\lambda_a = 1.3$, $\lambda_c = 2.5$, 15 A	162.50	159.37 (98.06)	166.84 (102.66)	7.47 (4.59)
	3,092	1,843 (59.43)	3,890 (125.44)	2,047 (66.01)
$\lambda_a = 1.3$, $\lambda_c = 2.5$, 15 A (2% CO)	163.29	160.17 (98.08)	166.54 (101.98)	6.37 (3.90)
	3,092	1,931 (62.50)	4,277 (138.44)	2,346 (75.93)
$\lambda_a = 1.3$, $\lambda_c = 2.5$, 15 A (4% CO)	163.62	161.30 (98.57)	165.90 (101.39)	4.60 (2.81)
	3,092	1,645 (53.04)	4,479 (144.43)	2,834 (91.38)
$\lambda_a = 1.3$, $\lambda_c = 2.5$, 15 A (6% CO)	164.27	161.64 (98.39)	167.49 (101.95)	5.85 (3.56)
	3,092	1,334 (42.97)	4,702 (151.47)	3,368 (108.50)

8.3.4 Type II flow-field – Operation with CO enriched hydrogen and air

The results of these measurements are shown in Fig.63.

8.3.5 Type III flow-field – Operation with hydrogen and air

In Fig.56(a), a load current of 10 A is drawn from the cell. The shape of the current density is the same as that shown in Fig.49. The solid-phase temperature distribution is defined by the shape of the current density (Fig.56(b)). As the stoichiometric flow rates are increased to 2.6 (anode) and 5 (cathode), the shape of the distribution becomes flatter due to the enhanced oxygen availability (Fig.56(c)). For these high flow rates, the mean solid-phase temperature decreases and the influence of the air entering the cell becomes more obvious close to the cathode inlet (Fig.56(d)).

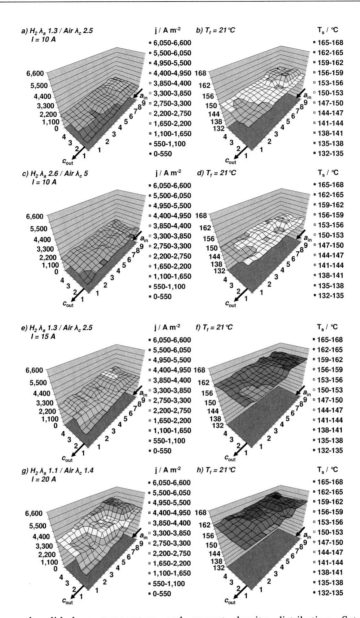

Fig.56. Measured solid-phase temperature and current density distribution. Set solid-phase temperature 150°C and 160°C and gas inlet temperature 21°C. Stoichiometric flow rates for 10 A (λ_a = 1.3, λ_c = 2.5 (a,b), λ_a = 2.6, λ_c = 5 (c,d)), 15 A (λ_a = 1.3, λ_c = 2.5 (e,f)), and 20 A load current (λ_a = 1.1, λ_c = 1.4 (g,h)).

Fig.56(e) presents the current density distribution for 15 A load current and stoichiometric flow rates of 1.3 (anode) and 2.5 (cathode). The solid-phase temperature distribution is highly similar to previous measurements. The air entering the cell at the cathode side only slightly influences the solid-phase temperature distribution close to the cathode inlet (Fig.56(f)). In Fig.56(g), the stoichiometric flow rates are reduced to 1.1 (anode) and 1.4 (cathode) while drawing 20 A from the cell. The oxygen availability dictates the shape of the current density distribution which is additionally overlapped by the hydrogen availability. The solid-phase temperature distribution in Fig.56(h) does not vary significantly from the previous measurements. The influence of the air entering the cell vanishes for low stoichiometric flow rates. Table 28 summarizes the results for the above measurements.

Table 28

Overview of measured solid-phase temperature and current density distribution for various operating conditions. Set solid-phase temperature and set current density equals 100%.

	Solid-phase temperature T_s and current density j			
	Mean	Minimum	Maximum	Difference
	°C	°C (%)	°C (%)	°C (%)
	A m^{-2}	A m^{-2} (%)	A m^{-2} (%)	A m^{-2} (%)
Measurement				
$\lambda_a = 1.3$, $\lambda_c = 2.5$, 10 A	151.48	148.30 (97.90)	156.28 (103.17)	7.89 (5.26)
	2,061	1,309 (63.20)	2,688 (129.79)	1,379 (66.58)
$\lambda_a = 2.6$, $\lambda_c = 5$, 10 A	150.54	146.68 (97.43)	153.65 (102.07)	6.97 (4.63)
	2,061	1,426 (68.91)	2,486 (120.14)	1,060 (51.22)
$\lambda_a = 1.3$, $\lambda_c = 2.5$, 15 A	160.92	158.09 (98.24)	164.23 (102.06)	6.14 (3.81)
	3,092	2,020 (65.28)	3,811 (123.17)	1,791 (57.88)
$\lambda_a = 1.1$, $\lambda_c = 1.4$, 20 A	161.66	158.04 (97.75)	165.43 (102.33)	7.39 (4.57)
	4,123	2,286 (55.31)	5,574 (134.88)	3,288 (79.56)

8.3.6 Type III flow-field – Operation with CO enriched hydrogen and air

Fig.57 presents the solid-phase temperature and current density distribution when drawing 20 A load current at stoichiometric flow rates of 1.3 (anode) and 2.5 (cathode). The distribution in Fig.57(a) and Fig.57(b) were recorded as reference measurements before increasing the CO content in the hydrogen to 2%, 4%, and 6%. For the solid-phase temperature, the shape is similar to previous results, whereas the influence of the air entering the cell at the cathode inlet becomes slightly more pronounced. A current density shift is observed when CO enriched hydrogen is used.

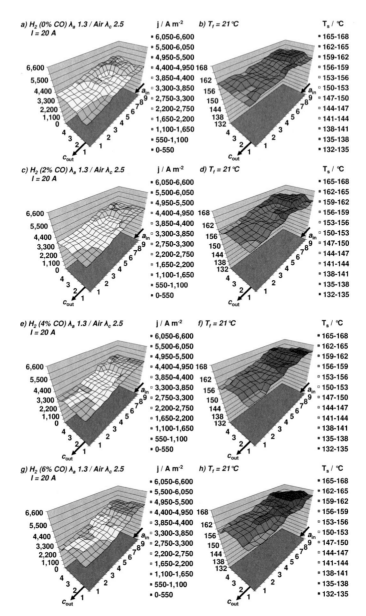

Fig.57. Measured solid-phase temperature and current density distribution. Set solid-phase temperature 160°C and gas inlet temperature 21°C. Stoichiometric flow rates for 20 A load current (0% CO (a,b), 2% CO (c,d), 4% CO (e,f), and 6% CO enriched hydrogen (g,h)).

The current density increases in the region of the anode inlet and decreases in the region of the anode outlet. This current density shift overlaps the distribution resulting from the oxygen availability (Fig.57(c)). The solid-phase temperature distribution in Fig.57(d) follows the current density distribution. A high overall solid-phase temperature is observed in the regions towards both inlets. As the CO content is increased to 4% and 6% the current density continuously shifts from the anode outlet to the region of the anode inlet. The hydrogen availability dictates the shape of the distribution. The highest current density values occur in the region of the anode inlet (Fig.57(e)), and the solid-phase temperature distribution follows the current density distribution, resulting in high overall values in the region of the anode inlet (Fig.57(f)). For 6% CO in the hydrogen, the current density gradient becomes large (Fig.57(g)). In this case, the highest values are found at the anode inlet, and the lowest values arise at the anode outlet. Under these operating conditions, a high overall solid-phase temperature is found close to the anode inlet while a lower temperature is observed towards both outlets, as shown in Fig.57(h). Table 29 summarizes the measured results. In this case, the difference is higher than that of the counter-flow configuration. The difference between the lowest and highest current density values increases with increasing CO content in the hydrogen.

Table 29

Overview of measured solid-phase temperature and current density distribution for various operating conditions. Set solid-phase temperature and set current density equals 100%.

	Solid-phase temperature T_s and current density j			
	Mean	Minimum	Maximum	Difference
	°C	°C (%)	°C (%)	°C (%)
	A m^{-2}	A m^{-2} (%)	A m^{-2} (%)	A m^{-2} (%)
Measurement				
$\lambda_a = 1.3$, $\lambda_c = 2.5$, 20 A	160.90	155.92 (96.90)	163.89 (101.86)	7.97 (4.95)
	4,123	3,043 (73.63)	4,965 (120.15)	1,922 (46.51)
$\lambda_a = 1.3$, $\lambda_c = 2.5$, 20 A (2% CO)	161.29	156.42 (96.98)	164.67 (102.10)	8.25 (5.11)
	4,123	3,423 (82.84)	4,845 (117.26)	1,422 (34.41)
$\lambda_a = 1.3$, $\lambda_c = 2.5$, 20 A (4% CO)	161.94	156.67 (96.74)	165.21 (102.02)	8.54 (5.27)
	4,123	3,393 (82.27)	4,915 (119.18)	1,522 (36.90)
$\lambda_a = 1.3$, $\lambda_c = 2.5$, 20 A (6% CO)	162.20	157.17 (96.89)	166.71 (102.78)	9.54 (5.88)
	4,123	2,995 (72.42)	5,202 (125.8)	2,207 (53.37)

9. Segmented EIS measurements in a HTPEM fuel cell

9.1 Type I flow-field

The cell is operated with hydrogen and air in counter-flow configuration at a set solid-phase temperature of 160°C while drawing 15 A load current. Fig.58 illustrates that the oxygen availability dictates the almost linear gradient from cathode inlet to cathode outlet. This result is comparable to the measured distribution shown in Fig.36(g). In total, 21 selected segments are sequentially scanned, and the data recorded. The light grey spectra represent the segments close to the cathode inlet, the medium grey spectra denote the segments in the middle over the membrane electrode assembly area, and the black spectra represent the segments towards the cathode outlet. The first intersection with the real axis occurs at 0.01 Ω cm² and is fairly constant for all spectra. The second intersection changes with the relative position of the segment over the membrane electrode assembly area. Table 30 summarizes selected current density values and fitted EIS model parameters.

Table 30

Current density values and fitted EIS model parameters. Mean value equals 100%.

Parameter / unit	Mean	Minimum (%) (segment)	Maximum (%) (segment)
j / A m^{-2}	3,092	1,550 (50.09) (S36)	4,423 (142.96) (S1)
R_{HF} / mΩ cm²	6.10	4.80 (78.77) (S9)	7.74 (122.55) (S22)
R_{LF} / mΩ cm²	22.25	15.57 (69.99) (S1)	28.57 (128.44) (S36)

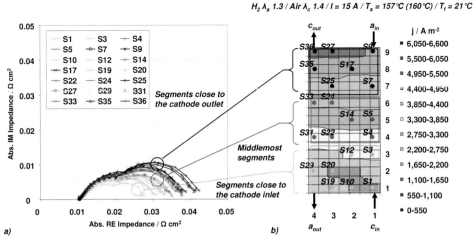

Fig.58. Local spectra obtained from segmented EIS measurements (a) and the measured current density

distribution (b). Set solid-phase temperature 160°C (157°C) and air inlet temperature 21°C. Stoichiometric flow rates for 15 A load current ($\lambda_a = 1.3$, $\lambda_c = 1.4$).

The spectra display a high frequency loop and a low frequency loop. Both loops become more visible as oxygen is continuously consumed towards the cathode outlet. Because the stoichiometric flow rates at the cathode side are relatively low, the low frequency loop may arise from oscillations in the gas partial pressure as suggested in [203,204]. As stated in [185], at low frequencies the oscillations in the reactant concentrations due to changes in the drawn load current extend into the gas channel. The oscillations are then compounded throughout the flow-field, resulting in a low frequency arc. This effect may be more pronounced for segments that are located closer to the cathode outlet.

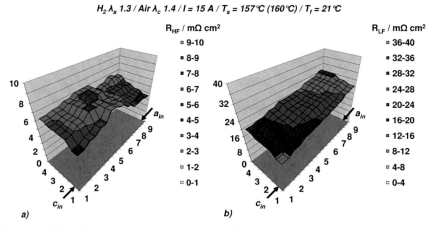

Fig.59. Fitted EIS model parameters for the high (a) and low frequency resistance (b) for the local spectra displayed in Fig.58(a). Set solid-phase temperature 160°C (157°C) and air inlet temperature 21°C. Stoichiometric flow rates for 15 A load current ($\lambda_a = 1.3$, $\lambda_c = 1.4$).

Fig.59 presents the fitted EIS model parameters for the spectra shown in Fig.58(b). The high frequency resistance slightly changes over the membrane electrode assembly area. The low frequency resistance increases continuously from cathode inlet to cathode outlet. These results are in agreement with the current density measurements shown in Fig.58(b). The Nernst resistance increases almost linearly from cathode inlet to cathode outlet. The low frequency capacity varies between 1.84 F and 2.59 F over the membrane electrode assembly area. The high frequency capacity varies between 5.11 F and 7.18 F over the membrane electrode assembly area and appears to increase slightly towards the anode inlet.

9.2 Type II flow-field

The measured current density distributions for the 4 stoichiometric flow rates and co-flow configurations while drawing 20 A load current at 160°C are depicted in Fig.60. The cell is operated with hydrogen and air. Very similar distributions are depicted in Fig.55(a) and Fig.55(c).

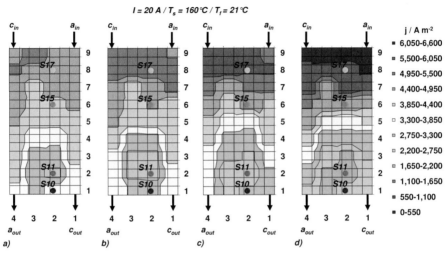

Fig.60. Measured current density distribution. Set solid-phase temperature 160°C and gas inlet temperature 21°C. Stoichiometric flow rates for 20 A load current ($\lambda_a = 1.3$, $\lambda_c = 4$ (a), $\lambda_a = 1.3$, $\lambda_c = 2.5$ (b), $\lambda_a = 1.3$, $\lambda_c = 2$ (c), $\lambda_a = 1.1$, $\lambda_c = 1.4$ (d)).

The highest current density is observed towards the cathode inlet and anode inlet. The shape of the current density distribution is defined by the poor fluid-flow distribution in the middlemost gas channel region and the oxygen availability. The lowest values over the membrane electrode assembly area appear near the cathode outlet. For all four stoichiometric flow rates, the current density distribution is higher on the left side over the membrane electrode assembly area compared to the right side. One possible explanation is that oxygen is more available in this region due to the number of straight channels in direct contact with the cathode inlet. As the stoichiometric flow rates are reduced, the current density distribution becomes more inhomogeneous, but the shape remains the same. Table 31 summarizes the influence of the stoichiometric flow rates at the anode and cathode side on the current density values.

Table 31

Current density values j. Mean value equals 100%

Stoichiometry	Mean	Minimum	Maximum
	A m^{-2}	A m^{-2} (%) (segment)	A m^{-2} (%) (segment)
$\lambda_a = 1.3, \lambda_c = 4$	4,123	2,707 (65.68) (S11)	5,327 (129.26) (S27)
$\lambda_a = 1.3, \lambda_c = 2.5$	4,123	2,577 (62,60) (S11)	5,476 (133.02) (S35)
$\lambda_a = 1.3, \lambda_c = 2$	4,123	2,227 (54.03) (S20)	5,936 (144.03) (S27)
$\lambda_a = 1.1, \lambda_c = 1.4$	4,123	1,916 (46.42) (S11)	6,566 (159.08) (S27)

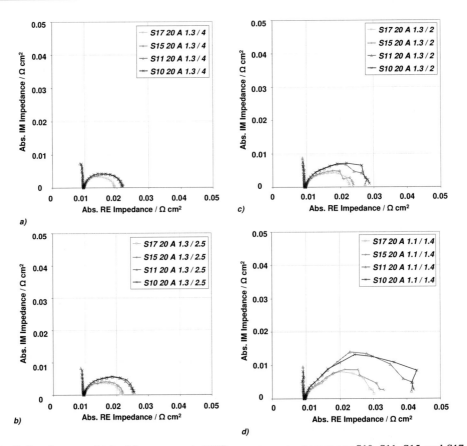

Fig.61. Local spectra obtained from segmented EIS measurements at segments S10, S11, S15, and S17. Set solid-phase temperature 160°C and air inlet temperature 21°C. Stoichiometric flow rates for 20 A load current ($\lambda_a = 1.3, \lambda_c = 4$ (a), $\lambda_a = 1.3, \lambda_c = 2.5$ (b), $\lambda_a = 1.3, \lambda_c = 2$ (c); $\lambda_a = 1.1, \lambda_c = 1.4$ (d)).

As can be seen from Fig.61, the first intersection of the RE axis occurs at 0.01 Ω cm^2 and is comparable to the value for type I flow-field. At stoichiometric flow rates of 1.3 (anode) and 4.0 (cathode) (Fig.61(a)), the local spectrum of segment S17 displays the lowest charge and gas transfer resistance, followed by the local spectra of segments S15, S11, and S10, which are similar in same size and shape. At stoichiometric flow rates of 1.3 (anode) and 2.5 (cathode), the local spectra for the segments differ, but their shapes remain similar (Fig.61(b)). The local spectrum for segment S17 exhibits the lowest charge and gas transfer resistances followed by segments S15, S11, and S10. It is remarkable that the local spectra for segments S17 and S15 appear similar in size and shape (as is the case for segments S11 and S10). The highest current density values were located near the cathode inlet, and the lowest value was observed in the middlemost gas channel region close to the cathode outlet. Reducing the stoichiometric flow rates to 1.3 (anode) and 2.0 (cathode) increased the current density gradient. Again, the local spectra for segments S17 and S15 are similar (Fig.61(c)). The local spectrum of segment S17 exhibits the lowest charge and gas transfer resistance, directly followed by segment S15. The local spectra of segments S11 and S10 also appear to be similar in shape but are significantly larger in size. Operating the cell at low stoichiometric flow rates further increases the charge and gas transfer resistances as the current density gradient increases (Fig.61(d)). The highest values are found at the cathode inlet, and the lowest values are found at the same position reported for the previous measurements. The local spectrum of segment S17 displays the lowest charge and gas transfer resistance, followed by segment S15. The local spectra of segments S11 and S10 again appear similar in size and shape but are significantly larger than those of segment S15 and S17.

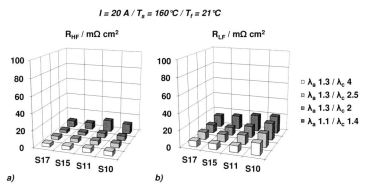

Fig.62. Fitted EIS model parameters for the high (a) and low frequency resistance (b) for the local spectra of the segments S10, S11, S15, and S17 shown in Fig.61. Set solid-phase temperature 160°C and air inlet temperature 21°C. Stoichiometric flow rates for 20 A load current.

The fitted EIS model parameters for the high and low frequency resistance shown in Fig.62 agree with the current density measurements and local spectra data. The high and low frequency resistance increase for the segments located near both outlets. Reducing the stoichiometric flow rate at the cathode side leads to an increase of the low frequency loop. The highest values are found for segments S10 and S11 at low stoichiometric flow rates of 1.1 (anode) and 1.4 (cathode). For these measurements, the Nernst resistance also increases for segments located near both outlets at reduced stoichiometric flow rates. The high frequency capacity increases slightly under reduced stoichiometric flow rates for all 4 segments, whereas the low frequency capacity decreases slightly for all 4 segments. Thus, the fluid-flow distribution and oxygen availability exert a major influence on the local parameters.

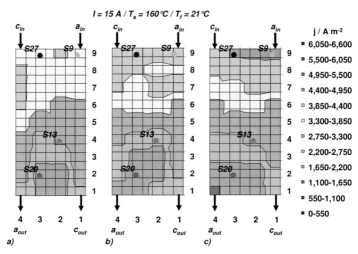

Fig.63. Measured current density distribution. Set solid-phase temperature 160°C and gas inlet temperature 21°C. Stoichiometric flow rates for 15 A load current (0% CO (a), 2% CO (b), and 4% CO enriched hydrogen (c)).

Fig.63(a) shows the results for a cell operating with CO enriched hydrogen and air at stoichiometric flow rates of 1.3 (anode) and 2.5 (cathode) when drawing 15 A load current (co-flow configuration). The set solid-phase temperature is 160°C. The current density distribution is similar to the distribution discussed above for the same stoichiometric flow rates. Fig.63(b) and Fig.63(c) depict the current density distribution for 15 A load current when operating the HTPEM fuel cell with CO enriched hydrogen. The shape of the current density imposed by the oxygen availability was overlapped by the hydrogen availability. As the CO content was increased, the current density decreased near the anode outlet and increased near the anode inlet. The distribution for 2% CO in the hydrogen yielded high values on the right side over the membrane electrode assembly area near the anode inlet. The lowest values were observed in the middlemost gas channel region towards the anode outlet. Fig.63(c) presents the results for 4% CO in the hydrogen. The oxygen and hydrogen availabilities now dictated the shape of the distribution. Under these operating conditions, the highest values are observed on the right side over the membrane electrode assembly area and especially in the anode inlet region. In fact, the current density increases close to the anode inlet and decreases close to the anode outlet. Table 32 provides selected current density values.

Table 32

Current density values j. Mean value equals 100%.

CO content	Mean	Minimum	Maximum
	A m^{-2}	A m^{-2} (%) (segment)	A m^{-2} (%) (segment)
0%	3,092	1,843 (59.43) (S11)	3,890 (125.44) (S18)
2%	3,092	1,931 (62.51) (S20)	4,277 (138.44) (S27)
4%	3,092	1,645 (53.04) (S28)	4,479 (144.43) (S18)

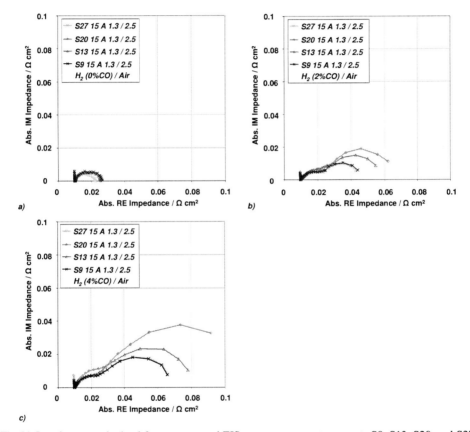

a)

b)

c)

Fig.64. Local spectra obtained from segmented EIS measurements at segments S9, S13, S20, and S27. Set solid-phase temperature 160°C and air inlet temperature 21°C. Stoichiometric flow rates for 15 A load current (0% CO (a), 2% CO (b), and 4% CO enriched hydrogen (c)).

Segment S27 is located near the cathode inlet, and the charge and gas transfer resistances at this point show the lowest value. The local spectra of the segments located near the cathode outlet exhibit slightly larger charge and gas transfer resistances (Fig.64(a)). Increasing the CO content in the hydrogen to 2% using the same stoichiometric flow rates changes the size and shape of all 4 local spectra as shown in Fig.64(b). The two loops can be clearly distinguished in the Nyquist plots. The local spectrum for segment S27 is similar to that of segment S9. The highest charge and gas transfer resistances are observed at segment S20 because this segment is closest to the anode outlet. A higher CO content in the hydrogen further increases the size of the local spectra (Fig.64(c)). Similar to previous observations, the local spectrum for segment S27 is almost identical to the local spectrum of segment S9. The local spectra of segment S20 and segment S13 are much larger in shape and size, with the second loop being dominant.

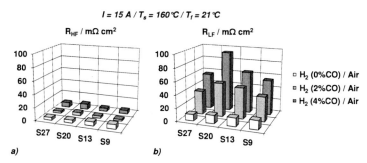

Fig.65. Fitted EIS model parameters for the high (a) and low frequency resistance (b) for the local spectra of segments S9, S13, S20, and S27 in Fig.64. Set solid-phase temperature 160°C and air inlet temperature 21°C. Stoichiometric flow rates for 15 A load current.

The fitted EIS model parameter values agree with all of the measurements (Fig.65). The high frequency resistance is nearly constant. The low frequency resistance increases as soon as the CO enriched hydrogen is used. The highest value is found for segment S20 followed by segment S13. The low frequency resistance has high values when using CO enriched hydrogen, especially at segments S20 and S13. As stated in [185], one possible explanation is, that the active sites within the reaction layer and especially in along the channel direction are occupied by CO. Hydrogen must diffuse farther before reaching a free site, which increases the local hydrogen concentration amplitudes and influences the current amplitude. Similar to the local oscillations in the oxygen partial pressure at the cathode side, it is possible that local oscillations in the hydrogen partial pressure at the anode side increase the low frequency resistance. This effect is more pronounced near the anode outlet because the hydrogen is continuously consumed. The model parameter fitting indicates that the Nernst resistance increases in a similar manner as the low frequency resistance does. For all 4 segments, the high frequency capacity increases for 2% CO in the hydrogen and decreases for 4% CO. The low frequency capacity of the 4 segments continuously decreases with higher CO content in the hydrogen.

9.3 Type III flow-field

The cell with type III flow-field is operated using hydrogen and air at 150°C while drawing 20 A load current. The stoichiometric flow rates are 1.3 (anode) and 2 (cathode) (Fig.66). The current density distribution shows the typical shape and characteristics dictated by the oxygen availability as observed previously. A similar shape is depicted in Fig.56(e).

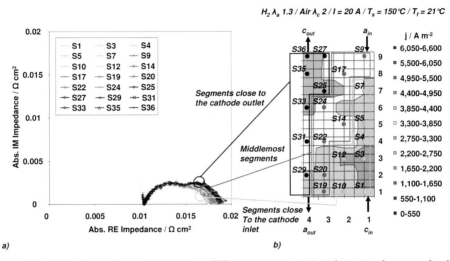

a)   b)

Fig.66. Local spectra obtained from segmented EIS measurements (a) and measured current density distribution (b). Set solid-phase temperature 150°C and air inlet temperature 21°C. Stoichiometric flow rates for 20 A load current ($\lambda_a = 1.3$, $\lambda_c = 2$).

Table 33 summarizes selected current density values and fitted EIS model parameters for the measurements shown in Fig.66.

Table 33
Current density values and fitted EIS model parameters. Mean value equals 100%.

Parameter / unit	Mean	Minimum (%) (segment)	Maximum (%) (segment)
j / A m^{-2}	4,123	3,044 (73.45) (S36)	5,106 (123.21) (S2)
R_{HF} / mΩ cm^2	3.32	3.13 (94.42) (S4)	3.45 (104.11) (S16)
R_{LF} / mΩ cm^2	5.88	5.04 (85.74) (S3)	7.09 (120.49) (S36)

Fig.66(b) depicts the local spectra of the selected segments. The first intersection with the real axis is somewhat above 0.01 Ω cm^2. This value is nearly constant and is comparable to previous measurements.

The second intersection changes with the position of the segment. Segments S1 and segment S2 are located at the cathode inlet and exhibit a reduced charge and gas frequency resistance. As oxygen was consumed towards the cathode outlet, both transfer resistances increased, and the two loops became slightly more visible. Thus, the local spectrum of segment S36 exhibits the highest charge and gas transfer resistance. Fig.67 shows the fitted EIS model parameters for these spectra. The high frequency resistance remains fairly constant over the membrane electrode assembly area. The low frequency resistance increases continuously from cathode inlet to cathode outlet. These results are in agreement with the current density measurements shown in Fig.66(a).

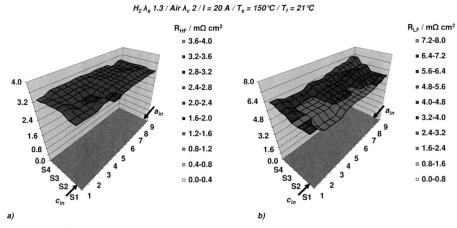

a)   b)

Fig.67. Fitted EIS model parameters for the high (a) and low frequency resistance (b) for the local spectra in Fig.66(a). Set solid-phase temperature 150°C and air inlet temperature 21°C. Stoichiometric flow rates for 20 A load current ($\lambda_a = 1.3$, $\lambda_c = 2$).

For the other parameters, the Nernst resistance follows the low frequency resistance, and the lowest values are obtained close to the cathode inlet, with the highest values occurring near the cathode outlet. The high frequency capacity is almost constant over the membrane electrode assembly area. The low frequency capacity decreases towards the cathode outlet. The highest value occurs near the cathode inlet whereas the lowest value is observed at the cathode outlet.

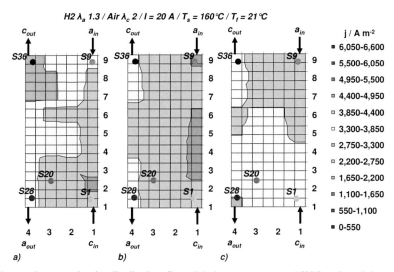

Fig.68. Measured current density distribution. Set solid-phase temperature 160°C and gas inlet temperature 21°C. Stoichiometric flow rates for 20 A load current (0% CO (a), 2% CO (b), and 4% CO enriched hydrogen (c)).

Fig.68(a) depicts the results for type III flow-field when drawing 20 A load current from the cell using stoichiometric flow rates of 1.3 (anode) and 2.5 (cathode). The set solid-phase temperature is 160°C. The current density distribution is very similar to the distribution observed in Fig.66(a). Table 34 provides selected current density values for different CO contents in the hydrogen.

Table 34

Current density values j. Mean value equals 100%.

CO content	Mean value	Minimum	Maximum
	A m^{-2}	A m^{-2} (%) (segment)	A m^{-2} (%) (segment)
0%	4,123	3,374 (81.83) (S36)	4,906 (118.99) (S2)
2%	4,123	3,681 (89.27) (S30)	4,731 (114.74) (S4)
4%	4,123	3,571 (86.61) (S28)	4,634 (112.39) (S7)

In the segmented EIS measurements, the different local spectra are all similar in size and shape when no CO is present in the hydrogen. A closer examination indicates that the local spectra of the segments closer to the cathode inlet are somewhat smaller in size and shape (Fig.69(a)). Increasing the CO content in the hydrogen to 2% changes the situation, with a second loop becoming visible (Fig.69(b)). The charge and gas transfer resistances are particularly higher for segment S28. Indeed, this segment is located at the anode outlet. Moreover, the charge and gas transfer resistances increase slightly in all local spectra. A similar current density shift from anode outlet to the anode inlet is observed as previously discussed for the other types of flow-fields. The local spectra for 4% CO in the hydrogen are shown in Fig.69(c). The charge and gas transfer resistances increase for all segments, and the second loop is clearly visible. Segment S9 is located at the anode inlet, and its local spectrum exhibits the lowest charge and gas transfer resistances. Similar to the values reported for 2% CO in the hydrogen, the current density decreases near the anode outlet and increases near the anode inlet.

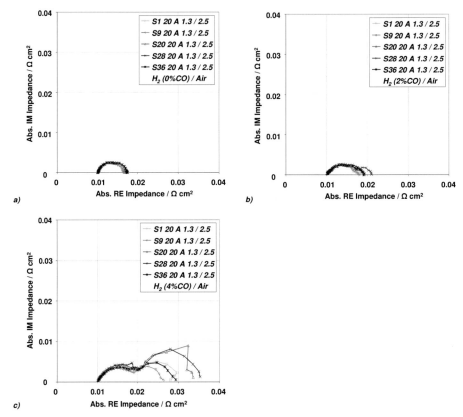

Fig.69. Local spectra obtained from segmented EIS measurements at segments S1, S9, S20, S28, and S36. Set solid-phase temperature 160°C and air inlet temperature 21°C. Stoichiometric flow rates for 20 A load current (0% CO (a), 2% CO (b), and 4% CO enriched hydrogen (c)).

Fig.70 presents the fitted EIS model parameters for the spectra shown in Fig.69. The high frequency resistance is fairly constant over the membrane electrode assembly area but appears to decrease slightly when increasing the CO content in the hydrogen. The low frequency resistance increases with increasing CO content. When no CO is present in the hydrogen, the low frequency resistance is more or less constant for all segments. When CO enriched hydrogen is used, the low frequency resistance increases for all segments especially for those near the anode outlet. These results are in agreement with the current density measurements shown in Fig.68. Local oscillations in the hydrogen partial pressure at the anode side could increase the low frequency resistance, being more pronounced for segments near the anode outlet.

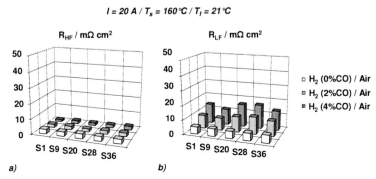

Fig.70. Fitted EIS model parameters for the high (a) and low frequency resistance (b) for the local spectra of segments S1, S9, S20, S28, and S36 in Fig.69. Set solid-phase temperature 160°C and air inlet temperature 21°C. Stoichiometric flow rates for 20 A load current.

10. Conclusion

10.1 Modeling and simulation

A complete three-dimensional model of a HTPEM fuel cell, employing a high temperature stable 50 cm^2 PBI/H_3PO_4 sol-gel membrane electrode assembly, was implemented, modeled, and solved using commercially available finite element software. The model provided a one-to-one representation of the HTPEM fuel cell that was used in all of the experimental tests. In total, three types of flow-fields were modeled and simulated at reference operating conditions. The developed model included the conservation equations of mass, momentum, species, charge, and energy, which were solved in their respective subdomains. A new two-phase temperature model was used to separately describe the fluid-(gas)-, and the solid-phase temperature distributions. An agglomerate model was employed to describe the electrochemical reactions. The presented model differs from previously published model as it separately accounts for the transport resistance of the species inside the spherical agglomerate, the amorphous phase phosphoric acid film, and the phosphoric acid water mixture film. The boundary equations were determined according to the experimental set-up. The modeling results were compared to experimental data, and similar trends were observed. Using a single set of modeling parameters, the model was able to predict the performance of type I (6 channel parallel serpentine flow-field) and type III flow-field (mixed-type flow-field) but failed to accurately predict the performance of type II flow-field (parallel straight flow-field) for a high current density. With respect to the modeling parameters, the transfer coefficient had significant impact on the results. The simulation results were evaluated and compared to segmented measurements, and similar trends were observed. Computational fluid dynamic results helped to support the observed trends in terms of the fluid-flow distribution and pressure drop for the three types of flow-fields. Generally spoken, the simulation results for type II flow-field indicated the most inhomogeneous fluid-flow distribution and oxygen availability throughout the flow-field, whereas the two other types of flow-fields exhibited a more homogenous distribution. The simulation results at no-load operating conditions, revealed an influence of the fluid-(gas)-phase temperature on the solid-phase temperature close to the cathode inlet. The variation ranged from 97% to 101% of the set solid-phase temperature. At load operating conditions, the current density gradients reported by the model were more pronounced than those reported by the segmented measurements. For all flow-fields, the model returned the highest current density values near the cathode inlet and the lowest near the cathode outlet. The highest solid-phase temperature occurred in the region of the highest current density. At reference operating conditions and 4,123 A m^{-2}, the current density varied between 35% and 160% for type II flow-field and between 52% and 144% for type I and type III flow-field. In contrast, the reported solid-phase temperature gradients were less pronounced than those reported by the segmented measurements and varied between 98% and 103%.

The simulation results also indicated that most of the heat is produced within the cathode side reaction layer. At reference operating conditions and 4,123 A m^{-2}, the highest overall solid-phase temperature is 3°C above the set solid-phase temperature. Most of the heat transfer between the two phases occurred via the cathode side gas channel walls.

10.2 Experimental

The nonsegmented HTPEM fuel cells were systematically characterized by first recording the pressure drop and the I-V-curve. At reference operating conditions, the performance of type I flow-field was the best (0.583 V at 4,123 A m^{-2}), directly followed by type III flow-field (0.574 V at 4,123 A m^{-2}). Type II flow-field exhibited a significantly lower overall performance (0.503 V at 4,123 A m^{-2}). The difference between co-flow and counter-flow configuration was minimal for type I and type III flow-field, whereas the difference was somewhat larger for type II flow-field. For the pressure drop, type II and type III flow-field exhibited the lowest values, whereas type I flow-field displayed the highest values. The HTPEM fuel cell with the three types of flow-fields displayed typical operational behaviour based on the stoichiometric flow rates, operating temperature, and CO enriched hydrogen. To gain more inside information during their operation, a segmented HTPEM fuel cell was developed and manufactured, and the solid-phase temperature and current density distribution were measured for the three types of flow-fields at various operating conditions (different load currents, stoichiometric flow rates, and anode gas compositions). The design of the segmented HTPEM fuel cell was similar to the original nonsegmented HTPEM fuel cell design. To perform segmented measurements, the bipolar-plate was segmented into 36 exchangeable segments and embedded in a high temperature stable polyetheretherketone matrix. This insulation material was also used at the anode side to minimize the influence of the heating elements on the solid-phase temperature distribution itself. The performance of the segmented HTPEM fuel cell was slightly below that of the nonsegmented HTPEM fuel cell due to a higher ohmic resistance of the complete set-up including the shunt resistance network. The solid-phase temperature and current density distribution were sequentially scanned, and the data were logged and displayed in a graphical user interface. The segmented cell was designed to also allow for segmented EIS measurements. These measurements were performed to qualitatively support the observed trends in the current density distributions for selected operating conditions.

10.3 Solid-phase temperature distribution

For the first time, solid-phase temperature distributions were measured for three types of flow-fields at no-load and load operating conditions. At no-load operating conditions, the distribution was uniform after the cell was heated-up to the set solid-phase temperature, which demonstrated the effectiveness of the polyetheretherketone insulation plates. When air entered the cell, an influence of the fluid-(gas)-phase temperature on the solid-phase temperature was observed for the three types of flow-fields (the same effect was observed when air entered the cell with a fluid-(gas)-phase temperature higher than the actual solid-phase temperature). For type I flow-field, a relatively large area over the membrane electrode assembly was cooled down by several degrees (5.38°C between the lowest and highest solid-phase temperature at 1 1 min^{-1}) because the fluid-flow was homogeneous within all gas channels. In type II flow-field the difference between the lowest and highest solid-phase temperature at 1 1 min^{-1} was 5.21°C and in type III flow-field it was 4.03°C. Thus, the shape of the flow-field, the fluid-flow distribution, and the flow rate are primarily responsible for a lower solid-phase temperature close to the cathode inlet. Overall, the highest solid-phase temperature was measured towards the outlet because heat accumulates in these regions. At load operating conditions, the highest solid-phase temperature was measured in the region of the highest current density, independent of the type of flow-field or flow configuration used. The measured distributions were in agreement with the simulations whereas the gradient in the measured distributions was smaller. The influence of the air entering the cell was relatively weak and was only noticed for high stoichiometric flow rates. For all flow-fields and most operating conditions, the solid-phase temperature varied between 95% and 102%, regardless of the flow configuration. During all tests, a higher overall solid-phase temperature was measured when CO enriched hydrogen was used compared to pure hydrogen. The reported findings provide valuable data for further nonisothermal modeling and simulation of HTPEM fuel cells and should increase our understanding of thermal stress and strain within the components of HTPEM fuel cells during operation. Moreover, the reported solid-phase temperature distributions should be of great interest for developing appropriate start-up and shut-down procedures for HTPEM fuel cells and stacks, particularly, for fuel cells with a significant difference between the fluid-(gas)-, and the solid-phase temperature. With respect to the solid-phase temperature distribution, the situation clearly changes for larger HTPEM fuel cell stacks. In this case, the high flow rates through the gas manifold, gas distributor, and flow-field should be analyzed. Additionally, such stacks are operated with external or internal forced cooling to operate it at the desired set solid-phase temperature. Finally, the interpretation of the results becomes more complicated as the structure of the flow-field becomes more complex. In this case, a finer segmentation could offer a clearer picture. Another possibility for improving the quality of the results would be the design of an ameliorated PID-controller for the heating elements to completely eliminate its minor influence on the overall solid-phase temperature (temperature overshoot).

10.4 Current density distribution

When the cell was operated with pure hydrogen, the shape of the current density was characterized by three interacting factors, namely, the oxygen availability, the shape of the flow-field, and the fluid-flow distribution. Under most operating conditions, the current density for the three types of flow-fields showed the highest values near the cathode inlet, with the lowest values occurring near the cathode outlet. For type II flow-field, the poor fluid-flow distribution significantly overlapped the current density distribution in the middlemost gas channel region. The set solid-phase temperature did not influence the shape of the current density distribution for the same stoichiometric flow rates. The almost linear gradient in the current density distribution from cathode inlet to cathode outlet increased as the stoichiometric flow rates were decreased at the cathode side. For typical stoichiometric flow rates, the current density varied between 70% and 120%. A high stoichiometric flow rate at the cathode side produced a relatively flat current density distribution. Consequently, a large gradient was observed from cathode inlet to cathode outlet for a low stoichiometric flow rate. For all types of flow-fields, the anode stoichiometric flow rates had a minor influence on the current density distribution up to values close to unity. The results changed when low stoichiometric flow rates were used on both sides of the cell. The oxygen availability still dictated the shape of the current density distribution, but the distribution was overlapped by the hydrogen availability. For the same operating conditions, the current density gradient was somewhat larger in co-flow than in counter-flow configuration. For typical stoichiometric flow rates, the current density varied between 56% and 135%. For some operating conditions, the current density at the cathode inlet was lower than the highest current density. The simulation results revealed that the phosphoric acid concentration, oxygen diffusion coefficient, and oxygen solubility vary in the region of the cathode inlet due to thermal interactions, which may cause this result. For all types of flow-fields, the segmented current density measurements displayed acceptable agreement with the simulations whereas the gradient in the measurements was not as pronounced. When CO enriched hydrogen was used, the cell voltage decreased as a constant power was drawn from the cell. The oxygen availability still defined the shape of the current density distribution for a low CO content in the hydrogen, but the current density increased in the region of the anode inlet and decreased in the region of the anode outlet. This effect was more pronounced as the amount of CO increased. In this case, the highest current density shifted towards the anode inlet and significantly overlapped the shape of the oxygen availability. The same behaviour was observed for all three types of flow-fields. The CO coverage increased towards the anode outlet because hydrogen was continuously consumed due to ongoing electrochemical reactions. A smaller fraction of hydrogen was supplied towards the anode outlet. A competitive adsorption mechanism between CO and hydrogen on the platinum surface occurred and determined the observed current density shift from anode outlet to anode inlet. The current density distribution switched back to its original shape after reducing the CO content in the hydrogen from several percent to 0%, meaning that the fuel cell performance was restored after poisoning. As previously mentioned, the solid-phase temperature distribution followed the current density shift when CO enriched hydrogen was used at the anode side. To conclude, when operating a HTPEM fuel cell in counter-flow configuration, the presence of small amounts of CO (e.g., 0.5-2%) beneficially overlaps the shape defined by the oxygen availability. In this case, the current density varied between 85% and 114%. When operating the HTPEM fuel cell in co-flow configuration the situation gets worse because the current density gradient from both inlets to both outlets increases. For a small amount of CO, the current density typically varied between 37% and 167%. From this perspective, it is recommended to operate a HTPEM fuel cell in counter-flow configuration and to design flow-fields with an anode outlet in the region of the cathode inlet and vice versa to minimize the current density gradient over the membrane electrode assembly area.

10.5 Segmented EIS measurements

For the first time and for selected operating conditions, segmented EIS measurements were acquired to support the observed current density distributions. The local spectra were evaluated using an equivalent circuit model to discuss the parameters. The high frequency resistance remained almost constant for the local spectra and did not change significantly with the position of the segment. The first intersection with the RE axis returned an ohmic resistance of approximately 0.01 Ω cm^2. Overall, the local spectra were somewhat larger in size for co-flow configuration than that for counter-flow configuration. The charge and gas transfer resistances displayed a significant dependence on the load current. The size and shape of the local spectrum decreased as the load current increased. When the cell was operated with hydrogen, the low frequency resistance increased for those segments near the cathode outlet especially for type I and type III flow-fields. For type II flow-field, the charge and gas transfer resistances also increased in regions with a poor fluid-flow distribution, as was the case in the middlemost gas channel region. For the three types of flow-fields, reducing the cathode stoichiometric flow rate increased both transfer resistances. A high cathode stoichiometric flow rate reduced both transfer resistances until a minimum was reached. For all of the flow-field types, a low cathode stoichiometric flow rate produced a second loop in the spectra, which may arise from oscillations in the oxygen partial pressure at the cathode side. The size and shape of the second loop increased with decreasing cathode stoichiometric flow rates, especially for the segments towards the cathode outlet. Variations in the anode stoichiometric flow rate did not measurably change the local spectra. Once CO enriched hydrogen was used, this situation changed. The local spectra increased significantly in size and shape, and a second loop appeared. This loop again increased when increasing the CO content in the hydrogen, especially for the segments towards the anode outlet. Along with the current density shift from anode outlet to anode inlet, for all three types of flow-fields, the highest charge and gas transfer resistances were observed for those segments near the anode outlet, whereas the lowest charge and gas transfer resistances were found for those near the anode inlet. The second loop may be related to CO adsorption at the platinum sites and to local oscillations in the hydrogen partial pressure at the anode side. The segmented EIS measurements should be confirmed by additional measurements using a more sophisticated set-up. The effect of the gas channel structure, the possible oscillations at low stoichiometric flow rates, and the influence of the porous media structure on the local spectra of a HTPEM fuel cell are interesting topics for future investigation. The present results demonstrate that a HTPEM fuel cell flow-field must be carefully designed. Moreover, the layout of the gas feeders, the positioning of the inlets and outlets and the co-flow and counter-flow configurations play an important role in ensuring homogeneous quantities distributions. With access to detailed inside information, one should be able to determine the weak points of a specific flow-field design and so to eliminate a high local current density peak or solid-phase temperature hot spots by using well designed components.

11. References

[1] Hydrogen and fuel cells – Fundamentals, technologies and applications. Stolten D. (editor), Wiley-VCH, Weinheim, Germany, 2010.

[2] http://www.calux.net (last accessed June 2012).

[3] http://enefield.eu (last accessed September 2013).

[4] Carter D.: Fuel cell residential micro-CHP developments in Japan. Fuel Cell Today Feb.29th, 2012 (2012).

[5] http://panasonic.co.jp/ap/FC/en_index.html (last accessed June 2012).

[6] http://www.baxi-innotech.de (last accessed June 2012).

[7] http://www.inhouse-engineering.de (last accessed June2012).

[8] The fuel cell today industry review 2012 – http://www.fuelcelltoday.com (last accessed February 2013).

[9] Wainright J.S., Wang J.T., Weng D., Savinell R.F., Litt M.: Acid-doped polybenzimidazoles: A new polymer electrolyte. J. Electrochem. Soc. 142 (1995), L121.

[10] Samms S.R., Wasmus S., Savinell R.F.: Thermal stability of proton conducting acid doped polybenzimidazole in simulated fuel cell environments. J. Electrochem. Soc. 143 (1996), 1225.

[11] Wang J.T., Savinell R.F., Wainright J.S., Litt M., Yu H.: A H_2/O_2 fuel cell using acid doped polybenzimidazole as a polymer electrolyte. Electrochim. Acta 41 (1996), 193.

[12] http://www.elcore2400.com (last accessed June 2012).

[13] http://www.clearedgepower.com (last accessed June 2012).

[14] Grove W.R.: On the gas voltaic battery. Experiments made with a view of ascertaining the rationale of its action and its application to eudiometry. Phil. Trans. R. Soc. Lond. 133 (1843), 91.

[15] Webb K.R.: Sir William Robert Grove (1811-1896) and the origins of the fuel cell. J. R. Inst. Chem. 85 (1961), 291.

[16] The birth of the fuel cell, 1835-1845 including the first publication of the complete correspondence from 1839 to 1868 between Christian Friedrich Schoenbein (discoverer of the fuel cell effect) and William Robert Grove (inventor of the fuel cell). Bossel U., European Fuel Cell Forum, Oberrohrdorf, Switzerland, 2000.

[17] PEM fuel cells – Theory and practice. Barbir F., Elsevier Academic Press, Burlington, U.S.A., 2005.

[18] Fuel cell systems explained. Larminie J., Dicks A., John Wiley & Sons Ltd., West Sussex, England, 2003.

[19] Fuel cells: From fundamentals to applications. Srinivasan S., Springer, New York, U.S.A., 2006.

[20] Fuel cell fundamentals. O'Hayre R., Cha S.-W., Colella W., Prinz B.F., John Wiley & Sons Inc., New York, U.S.A., 2006.

[21] Kerres J.: State of art of membrane development. J. Membr. Sci. 185 (2001), 1.

[22] Smitha B., Sridhar S., Khan A.A.: Solid polymer electrolyte membranes for fuel cell applications – a review. J. Membr. Sci. 259 (2005), 10.

[23] Alberti G., Casciola M.: Composite membranes for medium temperature PEM fuel cells. Annu. Rev. Mater. Res. 33 (2003), 129.

[24] Roziere J., Jones D.J.: Non-fluorinated polymer materials for proton exchange membrane fuel cells. Annu. Rev. Mater. Res. 33 (2003), 503.

[25] Jones D.J., Roziere J.: Recent advances in the functionalization of polybenzimidazole and polyetherketone for fuel cell applications. J. Membr. Sci. 185 (2001), 41.

[26] Kreuer K.D.: On the development of proton conducting polymer membranes for hydrogen and methanol fuel cells. J. Membr. Sci. 185 (2001), 29.

[27] Lobato J., Cañizares P., Manuel A., Úbeda D., Pinar F.J.: Enhancement of the fuel cell performance of a high temperature PEMFC running with titanium composite polybenzimidazole-based membranes. J. Power Sources 196 (2010), 8265.

[28] Lobato J., Cañizares P., Rodrigo M.A., Úbeda D., Pinar F.J.: A novel titanium PBI-based composite membrane for high temperature PEMFCs. J. Membr. Sci. 369 (2011), 105.

[29] Kurdakova V., Quartarone E., Mustarelli P., Magistris A., Caponetti E., Saladino M.L.: PBI-based composite membranes for polymer fuel cells. J. Power Sources 195 (2010), 7765.

[30] Angioni A., Righetti P.P., Quartarone E., Dilena E., Mustarelli P., Magistris A.: Novel aryloxy-polybenzimidazoles as proton conducting membranes for high temperature PEMFCs. Int. J. Hydrogen Energy 36 (2011), 7174.

[31] Aili D., Hansen M.K., Pan C., Li Q., Christensen E., Jensen J.O., Bjerrum N.J.: Phosphoric acid doped membranes based on Nafion®, PBI and their blends – Membrane preparation, characterization and steam electrolysis testing. Int. J. Hydrogen Energy 36 (2011), 6985.

[32] Li Q.F., Rudbeck H.C., Chromik A., Jensen J.O., Pan C., Steenberg T., Calverley M., Bjerrum N.J., Kerres J.: Properties, degradation and high temperature fuel cell test of different types of PBI and PBI blend membranes. J. Membr. Sci. 347 (2010), 260.

[33] Savinell R., Yeager E., Tryk D., Landau U., Wainright J., Weng D., Lux K., Litt M., Rogers C.: A polymer electrolyte for operation at temperatures up to 200°C. J. Electrochem. Soc. 141 (1994), L46.

[34] Zhang J., Xie Z., Zhang J., Tang Y., Song C., Navessin T., Shi Z., Song D., Wang H., Wilkinson D.P., Liu Z.-S., Holdcroft S.: High temperature PEM fuel cells. J. Power Sources 160 (2006), 872.

[35] Li Q., He R., Jensen J.O., Bjerrum, N.J.: PBI-based polymer membranes for high temperature fuel cells – preparation, characterization and fuel cell operation. Fuel Cells 4 (2004), 147.

[36] Li Q., Jensen J.O., Savinell R.F., Bjerrum N.J.: High temperature proton exchange membranes based on polybenzimidazoles for fuel cells. Prog. Polym. Sci. 34 (2009), 449.

[37] Xiao L., Zhang H., Scanlon E., Ramanathan S., Choe E.-W., Rogers D., Apple T., Benicewicz B.C.: High-temperature polybenzimidazoles for fuel cell membranes via a sol-gel process. Chem. Mater. 17 (2005), 5328.

[38] http://www.fuel-cell.basf.com/ca/internet/Fuel_Cell/ (last accessed June 2012).

[39] http://www.fumatech.com (last accessed June 2012).

[40] http://www.adventech.gr/ (last accessed June 2012).

[41] Systematic characterization of HT PEMFCs containing PBI/H₃PO₄ systems. Bandlamudi G. (Dissertation), Logos Verlag, Berlin, Germany, 2011.

[42] Ma Y.-L., Wainright J.S., Litt M.H., Savinell R.F.: Conductivity of PBI membranes for high-temperature polymer electrolyte fuel cells. J. Electrochem. Soc. 151 (2004), A8.

[43] Schmidt T.J.: Durability and degradation in high-temperature polymer electrolyte fuel cells. ECS Trans. 1 (2006), 19.

[44] Schmidt T.J., Baurmeister J.: Properties of high-temperature PEFC Celtec®-P 1000 MEAs in start/stop operation mode. J. Power Sources 176 (2008), 428.

[45] Wippermann K., Wannek C., Oetjen H.F., Mergel J., Lehnert W.: Cell resistance of poly(2,5-benzimidazole)-based high temperature polymer membrane fuel cell membrane electrode assemblies: time dependence and influence of the operating parameters. J. Power Sources 195 (2010), 2806.

[46] Yu S., Xiao L., Benicewicz B.C.: Durability studies of PBI-based high temperature PEMFCs. Fuel Cells 8 (2008), 165.

[47] Gang X., Hjuler H.A., Olsen C., Berg R.W., Bjerrum N.J.: Electrolyte addaitives for phosphoric acid fuel cells. J. Electrochem. Soc. 140 (1993), 896.

[48] Hong S.-G., Kwon K., Lee M.-J., Yoo D.Y.: Performance enhancement of phosphroric acid-based proton exchange membrane fuel cells by using ammonium trifluoromethansulfonate. Electrochem. Commun. 11 (2009), 1124.

[49] Wannek C., Konradi I., Mergel J., Lehnert W.: Redistribution of phosphoric acid in membrane electrode assemblies for high-temperature polymer electrolyte fuel cells. Int. J. Hydrogen Energy 34 (2009), 9479.

[50] Kwon K., Park J.O., Yoo D.Y., Yi J.S.: Phosphoric acid distribution in the membrane electrode assembly of high temperature proton exchange membrane fuel cell. Electrochim. Acta 54 (2009), 6570.

[51] Scholta J., Kuhn R., Wazlawik S., Jörissen L.: Startup-procedures for a HT-PEMFC stack. ECS Trans. 17 (2009), 325.

[52] Scholta J., Messerschmidt M., Jörissen L., Hartnig C.: Externally cooled high temperature polymer electrolyte membrane fuel cell stack. J. Power Sources 190 (2009), 83.

[53] Scholta J., Zhang W., Jörissen L., Lehnert W.: Conceptual design for an externally cooled HT-PEMFC stack. ECS Trans. 12 (2008), 113.

[54] Andreasen S.J., Kær S.K.: Modelling and evaluation of heating strategies for high temperature polymer electrolyte membrane fuel cell stacks. Int. J. Hydrogen Energy 33 (2008), 4655.

[55] Jensen H.-C.B., Kær S.K.: Boundary model-based reference control of blower cooled high temperature polymer electrolyte membrane fuel cells. Int. J. Hydrogen Energy 36 (2011), 5030.

[56] http://www.zahner.de (last accessed June 2012).

[57] Impedance spectroscopy: Theory, experiment, and applications (2nd edition). Macdonald J.R., Barsoukov E., John Wiley & Sons Inc., Hoboken, U.S.A., 2005.

[58] Electrochemical impedance spectroscopy. Orazem M.E., Tribollet B., John Wiley & Sons Inc., Hoboken, U.S.A., 2008.

[59] Gomadam P.M., Weidner J.W.: Analysis of electrochemical impedance spectroscopy in proton exchange membrane fuel cells. Int. J. Energy Res. 29 (2005), 1133.

[60] Cheddie D., Munroe N.: Review and comparison of approaches to proton exchange membrane fuel cell modeling. J. Power Sources 147 (2005), 72.

[61] Haraldsson K., Wipke K.: Evaluating PEM fuel cell system models. J. Power Sources 126 (2004), 88.

[62] Djilali N.: Computational modelling of polymer electrolyte membrane (PEM) fuel cells: Challenges and opportunities. Energy 32 (2007), 269.

[63] Weber A.Z., Newman J.: Modeling transport in polymer-electrolyte fuel cells. Chem. Rev. 104 (2004), 4679.

[64] Wang C.Y.: Fundamental models for fuel cell engineering. Chem. Rev. 104 (2004), 4727.

[65] Yao K.Z., Karan K., McAuley K.B., Oosthuizen P., Peppley B., Xie T.: A review of mathematical models for hydrogen and direct methanol polymer electrolyte membrane fuel cells. Fuel Cells 4 (2004), 3.

[66] Bıyıkoğlu A.: Review of proton exchange membrane fuel cell models. Int. J. Hydrogen Energy 30 (2005), 1181.

[67] Faghri A., Guo Z.: Challenges and opportunities of thermal management issues related to fuel cell technology and modeling. Int. J. Heat Mass Transfer 48 (2005), 3891.

[68] Costamagna P., Srinivasan S.: Quantum jumps in the PEMFC science and technology from the 1960s to the year 2000: Part I. Fundamental scientific aspects. J. Power Sources 102 (2001), 242.

[69] Costamagna P., Srinivasan S.: Quantum jumps in the PEMFC science and technology from the 1960s to the year 2000: Part II. Engineering, technology development and application aspects. J. Power Sources 102 (2001), 253.

[70] Siegel C.: Review of computational heat and mass transfer modeling in polymer-electrolyte-membrane (PEM) fuel cells. Energy 33 (2008), 1331.

[71] http://www.comsol.com (last accessed June 2012).

[72] Transport phenomena in fuel cells – Series: Developments in heat transfer, Vol.19. Sundén B., Faghri M. (editors), WIT Press, Southampton, U.K., 2005.

[73] Korsgaard A.R., Refshauge R., Nielsen M.P., Bang M., Kær S.K.: Experimental characterization and modeling of commercial polybenzimidazole-based MEA performance. J. Power Sources 162 (2006), 239.

[74] Cheddie D., Munroe N.: Parametric model of an intermediate temperature PEMFC. J. Power Sources 156 (2006), 414.

[75] Cheddie D., Munroe N.: Three dimensional modeling of high temperature PEM fuel cells. J. Power Sources 160 (2006), 215.

[76] Cheddie D., Munroe N.: Mathematical model of a PEMFC using a PBI membrane. Energy Convers. Manage. 47 (2006), 1490.

[77] Cheddie D., Munroe N.: A two-phase model of an intermediate temperature PEMFC. Int. J. Hydrogen Energy 32 (2007), 832.

[78] Peng J., Lee S.J.: Numerical simulation of proton exchange membrane fuel cells at high operating temperature. J. Power Sources 162 (2006), 1182.

[79] Peng J., Lee S.J.: Transient response of high temperature PEM fuel cell. J. Power Sources 179 (2008), 220.

[80] Scott K., Pilditch S., Mamlouk M.: Modelling and experimental validation of a high temperature polymer electrolyte fuel cell. J. Appl. Electrochem. 37 (2007), 1245.

[81] Ubong E.U., Shi Z., Wang X.: A-3D-modeling and experimental validation of a high temperature PBI based PEMFC. ECS Trans. 16 (2008), 79.

[82] Simulation einer Hochtemperatur-PEM-Brennstoffzelle. Schaar B. (Dissertation), AutoUni – Schriftreihe, Logos Verlag, Berlin, Germany, 2008.

[83] Shamardina O., Chertovich A., Kulikovsky A.A., Khokhlov A.R.: A simple model of a high temperature PEM fuel cell. Int. J. Hydrogen Energy 35 (2010), 9954.

[84] Kulikovsky A.A., Oetjen H.-F., Wannek Ch.: A simple and accurate method for high-temperature PEM fuel cell characterisation. Fuel Cells 10 (2010), 363.

[85] Sousa T., Mamlouk M., Scott K.: An isothermal model of a laboratory intermediate temperature fuel cell using PBI doped phosphoric acid membranes. Chem. Eng. Sci. 65 (2010), 2513.

[86] Zenith F., Seland F., Kongstein O.E., Børresen B., Tunold R., Skogestad S.: Control-oriented modelling and experimental study of the transient response of a high-temperature polymer fuel cell. J. Power Sources 162 (2006), 215.

[87] Wang C.-P., Chu H.-S., Yan Y.-Y., Hsueh K.-L.: Transient evolution of carbon monoxide poisoning effect of PBI membrane fuel cells. J. Power Sources 170 (2007), 235.

[88] Bergmann A., Gerteisen D., Kurz T.: Modelling of CO poisoning and its dynamics in HTPEM, Fuel cells 10 (2010), 278.

[89] Siegel C., Bandlamudi G., Heinzel A.: Numerical simulation of a high-temperature PEM (HTPEM) fuel cell. In: Proceedings of the European COMSOL conference, Grenoble, France, 2007.

[90] Siegel C., Bandlamudi G., Heinzel A.: Modeling polybenzimidazole/phosphoric acid membrane behaviour in a HTPEM fuel cell. In: Proceedings of the European COMSOL conference, Hannover, Germany, 2008.

[91] Siegel C., Bandlamudi G., Heinzel A.: Large scale 3D flow distribution analysis in HTPEM fuel cells. In: Proceedings of the European COMSOL conference, Milan, Italy, 2009.

[92] Siegel C., Bandlamudi G., Heinzel A.: Evaluating the effects of stack compression on the physical characteristics of HT PEMFCs with CFD modelling software. In: Proceedings of the fuel cell science & technology conference, Copenhagen, Denmark, 2008.

[93] Siegel C., Bandlamudi G., Heinzel A.: Systematic characterization of a PBI/H_3PO_4 sol-gel membrane – Modeling and simulation. J. Power Sources 196 (2011), 2735.

[94] Kvesić M., Reimer U., Froning D., Lüke L., Lehnert W., Stolten D.: 3D modeling of a 200 cm^2 HT-PEFC short stack. Int. J. Hydrogen Energy 37 (2012), 2430.

[95] Lüke L., Janßen H., Kvesić M., Lehnert W., Stolten D.: Performance analysis of HT-PEFC stacks. Int. J. Hydrogen Energy 37 (2012), 9171.

[96] Doubek G., Robalinho E., Cunha E.F., Cekinski E., Linardi M.: Application of CFD techniques in the modelling and simulation of PBI PEMFC. Fuel cells 11 (2011), 764.

[97] Sousa T., Mamlouk M., Scott K.: A non-isothermal model of a laboratory intermediate temperature fuel cell using PBI doped phosphoric acid membranes. Fuel Cells 10 (2010), 993.

[98] Mamlouk M., Sousa T., Scott K.: A high temperature polymer electrolyte membrane fuel cell model for reformate gas. International Journal of Electrochemistry 2011 (2011), Article ID 520473.

[99] Entwicklung und Charakterisierung eines portable Hochtemperatur-PEM-Brennstoffzellensystems. Kurz T. (Dissertation), Fraunhofer Verlag, Stuttgart, Germany, 2011.

[100] Reddy E.H., Monder D.S., Jayanti S.: Parametric study of an external coolant system for a high temperature polymer electrolyte membrane fuel cell. Appl. Therm. Eng. 58 (2013), 155.

[101] Jiao K., Zhou Y., Du Q., Yin Y., Yu S., Li X.: Numerical simulations of carbon monoxide poisoning in high temperature proton exchange membrane fuel cells with various flow channel designs. Appl. Energ. 104 (2013), 21.

[102] Hwang J.J.: Thermal-electrochemical modelling porous electrodes of a PEM fuel cell. J. Electrochem. Soc. 153 (2006), A216.

[103] Epting W.K., Litster S.: Effects of an agglomerate size distribution on the PEFC agglomerate model. Int. J. Hydrogen Energy 37 (2012), 8505.

[104] Harvey D., Pharoah J.G., Karan K.: A comparison of different approaches to modelling the PEMFC catalyst layer. J. Power Sources 179 (2008), 209.

[105] Secanell M., Karan K., Suleman A., Djilali N.: Multi-variable optimization of PEMFC cathodes using an agglomerate model. Electrochim. Acta 52 (2007), 6318.

[106] Choudhury S.R., Deshmukh M.B., Rengaswamy R.: A two-dimensional steady-state model for phosphoric acid fuel cells (PAFC). J. Power Sources 112 (2002), 137.

[107] Characterization of the components of the proton exchange membrane fuel cell. Broka K. (Techn. Lic. Thesis), Royal Institute of Technology, Stockholm, Sweden, 1995.

[108] Sun W., Peppley B.A., Karan K.: An improved two-dimensional agglomerate cathode model to study the influence of catalyst layer structural parameters. Electrochim. Acta 50 (2005), 3359.

[109] Sasmito A.P., Mujumdar A.S.: Performance evaluation of a polymer electrolyte fuel cell with a dead-end anode: A computational fluid dynamic study. Int. J. Hydrogen Energy 36 (2011), 10917.

[110] Transport phenomena (2nd edition). Bird R.B., Stewart W.E., Lightfoot E.N., John Wiley & Sons Inc., New York, U.S.A., 2007.

[111] Liu Z., Wainright J.S., Litt M.H., Savinell R.F.: Study of the oxygen reduction reaction (ORR) at Pt interfaced with phosphoric acid doped polybenzimidazole at elevated temperature and low relative humidity. Electrochim. Acta 51 (2006), 3914.

[112] Computational modeling and optimization of proton exchange membrane fuel cells. Gallart M.S. (Dissertation), University of Victoria, Victoria, Canada, 2007.

[113] Li Q., Jensen J.O., Savinell R.F., Bjerrum N.J.: High temperature proton exchange membranes based on polybenzimidazoles for fuel cells. Prog. Polym. Sci. 34 (2009), 449.

[114] Li Q., Gang X., Hjuler H.A., Berg R.W., Bjerrum N.J.: Oxygen reduction on gas-diffusion electrodes for phosphoric acid fuel cells by a potential decay method. J. Electrochem. Soc. 142 (1995), 3250.

[115] Scharifker B.R., Zelenay P., Bockris J.O'M.: Kinetics of oxygen reduction in molten phosphoric acid at high temperatures. J. Electrochem. Soc. 134 (1987), 2714.

[116] Appleby A.J.: Evolution and reduction of oxygen on oxidized platinum in 85% orthophosphoric acid. J. Electroanal. Chem. Interfacial Electrochem. 24 (1970), 97.

[117] Appleby A.J.: Oxygen reduction on oxide-free platinum in 85% orthophosphoric acid: temperature and impurity dependence. J. Electrochem. Soc. 117 (1970), 328.

[118] Huang J.C., Sen R.K., Yeager E.: Oxygen reduction on platinum in 85% orthophosphoric acid. J. Electrochem. Soc. 126 (1979), 786.

[119] O'Grady W.E., Tayler E.J., Srinivasan S.: Electroreduction of oxygen on reduced platinum in 85% phosphoric acid. J. Electroanal. Chem. 132 (1982), 137.

[120] McBreen J., O'Grady W.E., Richter R.: A rotating disk electrode apparatus for the study of fuel cell reactions at elevated temperatures and pressures. J. Electrochem. Soc. 131 (1984), 1215.

[121] Bregoli L.J.: The influence of platinum crystallite size on the electrochemical reduction of oxygen in phosphoric acid. Electrochim. Acta 23 (1978), 489.

[122] Yang S.-C.: Modeling and simulation of steady-state polarization and impedance response of phosphoric acid fuel-cell cathodes with catalyst-layer microstructure consideration. J. Electrochem. Soc. 147 (2000), 71.

[123] Glass J.T., Cahen G.L., Stoner G.E.: The effect of phosphoric acid concentration on electrocatalysis. J. Electrochem. Soc. 136 (1989), 656.

[124] Hsueh K.-L., Gonzales E.R., Srinivasan S., Chin D.-T.: Effects of phosphoric acid concentration on oxygen reduction kinetics at platinum. J. Electrochem. Soc. 131 (1984), 823.

[125] Kunz H.R., Gruver G.A.: Catalytic activity of active platinum supported on carbon for electrochemical oxygen reduction in phosphoric acid. J. Electrochem. Soc. 122 (1975), 1279.

[126] Jalani N.H., Ramani M., Ohlsson K., Buelte S., Pacifico G., Pollard R., Staudt R., Datta R.: Performance analysis and impedance spectral signatures of high temperature PBI – phosphoric acid gel membrane fuel cells. J. Power Sources 160 (2006), 1096.

[127] Klinedinst K., Bett J.A.S., MacDonald J., Stonehart P.: Oxygen solubility and diffusivity in hot concentrated H_3PO_4. J. Electroanal. Chem. Interfacial Electrochem. 57 (1974), 281.

[128] Gan F., Chin D.-T.: Determination of diffusivity and solubility of oxygen in phosphoric acid using a transit time on a rotating ring-disc electrode. J. Appl. Electrochem. 23 (1993), 452.

[129] Roethlein R.J., Maget H.J.R.: The electrochemical reduction of oxygen on electrodes partially immersed in phosphoric. J. Electrochem. Soc. 113 (1966), 581.

[130] Gubbins K.E., Walker R.D.: The solubility and diffusivity of oxygen in electrolytic solutions. J. Electrochem. Soc. 112 (1965), 469.

[131] Liu Z., Wainright J.S., Savinell R.F.: High-temperature polymer electrolytes for PEM fuel cells: study of the oxygen reduction reaction (ORR) at a Pt-polymer electrolyte interface. Chem. Eng. Sci. 59 (2004), 4833.

[132] MacDonald D.I., Boyack J.R.: Density, electrical conductivity, and vapor pressure of concentrated phosphoric acid. Chem. Eng. Data 14 (1969), 380.

[133] The fundamental studies of polybenzimidazole/phosphoric acid polymer electrolyte for fuel cells. Ma Y. (Dissertation), Case Western Reserve University, Cleveland, U.S.A., 2004.

[134] Zhang J., Tang Y., Song C., Zhang J.: Polybenzimidazole-membrane-based PEM fuel cell in the temperature range of 120–200°C. J. Power Sources 172 (2007), 163.

[135] http://www.etek-inc.com (last accessed June 2009).

[136] Rao R.M., Bhattacharyya D., Rengaswamy R., Choudhury S.R.: A two-dimensional steady state model including the effect of liquid water for a PEM fuel cell cathode. J. Power Sources 173 (2007), 375.

[137] Turnbull A.G.: Thermal conductivity of phosphoric acid. J. Chem. Eng. Data 10 (1965), 118.

[138] http://www.eisenhuth.de (last accessed June 2012).

[139] Cindrella L., Kannan A.M., Lin J.F., Saminathan K., Ho Y., Lin C.W., Wertz J.: Gas diffusion layer for proton exchange membrane fuel cells – A review. J. Power Sources 194 (2009), 146.

[140] Inhomogeneous compression of PEMFC gas diffusion layers. Nitta I. (Dissertation), Helsinki University of Technology, Helsinki, Finland, 2008.

[141] Antolini E.: Carbon supports for low-temperature fuel cell catalysts. Appl. Catal. B: Environmental 88 (2009), 1.

[142] Mass transfer in multicomponent mixtures. Wesselingh J.A., Krishna R., Delft University Press, Delft, Netherlands, 2000.

[143] Khandelwal M., Mench M.M.: Direct measurement of through-plane thermal conductivity and contact resistance in fuel cell materials. J. Power Sources 161 (2006), 1106.

[144] Tüber K., Pócza D., Hebling C.: Visualization of water buildup in the cathode of a transparent PEM fuel cell. J. Power Sources 124 (2003), 403.

[145] Zhang F.Y., Yang X.G., Wang C.Y.: Liquid water removal from a polymer electrolyte fuel cell. J. Electrochem. Soc. 153 (2006), A225.

[146] Nishida K., Murakami T., Tsushima S., Hirai S.: Measurement of liquid water content in cathode gas diffusion electrode of polymer electrolyte fuel cell. J. Power Sources 195 (2010), 3365.

[147] Turhan A., Heller K., Brenizer J.S., Mench M.M.: Quantification of liquid water accumulation and distribution in a polymer electrolyte fuel cell using neutron imaging. J. Power Sources 160 (2006), 1195.

[148] Hickner M.A., Siegel N.P., Chen K.S., McBrayer D.N., Hussey D.S., Jacobson D.L., Arif M.: Real-time imaging of liquid water in an operating proton exchange membrane fuel cell. J. Electrochem. Soc. 153 (2006), A902.

[149] Yoshizawa K., Ikezoe K., Tasaki Y., Kramer D., Lehmann E.H., Scherer G.G.: Analysis of gas diffusion layer and flow-field design in a PEMFC using neutron radiography. J. Electrochem. Soc. 155 (2008), B223.

[150] Satija R., Jacobson D.L., Arif M., Werner S.A.: In situ neutron imaging technique for evaluation of water management systems in operating PEM fuel cells. J. Power Sources 129 (2004), 238.

[151] Boillat P., Kramer D., Seyfang B.C., Frei G., Lehmann E., Scherer G.G., Wokaun A., Ichikawa Y., Tasaki Y., Shinohara K.: In situ observation of the water distribution across a PEFC using high resultion neutron radiography. Electrochem. Commun. 10 (2008), 546.

[152] Sasabe T., Tsushima S., Hirai S.: In-situ visualization of liquid water in an operating PEMFC by soft X-ray radiography. Int. J. Hydrogen Energy 35 (2010), 11119.

[153] Kuhn R., Scholta J., Krüger Ph., Hartnig Ch., Lehnert W., Arlt T., Manke I.: Measuring device for synchrotron X-ray imaging and first results of high temperature polymer electrolyte membrane fuel cells. J. Power Sources 196 (2011), 5231.

[154] Arlt T., Maier W., Tötzke C., Wannek C., Markötter H., Wieder F., Banhart J., Lehnert W., Manke I.: Synchrotron X-ray radioscopic in situ study of high-temperature polymer electrolyte fuel cells – Effect of operation conditions on structure of membrane. J. Power Sources 246 (2014), 290.

[155] Shim J.Y., Tsushima S., Hirai S.: High resolution MRI investigation of transversal water content distributions in PEM under fuel cell operation. ECS Trans. 25 (2009), 523.

[156] Andreasen S.J., Kær S.K.: Modelling and evaluation of heating strategies for high temperature polymer electrolyte membrane fuel cell stacks. Int. J. Hydrogen Energy 33 (2008), 4655.

[157] Jensen H.-C.B., Kær S.K.: Boundary model-based reference control of blower cooled high temperature polymer electrolyte membrane fuel cells. Int. J. Hydrogen Energy 36 (2011), 5030.

[158] Mench M., Burford D., Davis T.: In situ temperature distribution measurements in an operating polymer electrolyte fuel cell. In: Proceedings of IMECE'03, Washington D.C., U.S.A., 2003.

[159] Wilkinson M., Blanco M., Gu E., Martin J.J., Wilkinson D.P., Zhang J.J., Wang H.: In situ experimental technique for measurement of temperature and current distribution in proton exchange membrane fuel cells. Electrochem. Solid State Lett. 9 (2006), A507.

[160] He S., Mench M.M., Tadigadapa S.: Thin film temperature sensor for real-time measurement of electrolyte temperature in a polymer electrolyte fuel cell. Sensor. Actuator. – A Phys. 125 (2006), 170.

[161] Wang M., Guo H., Ma C.: Temperature distribution on the MEA surface of a PEMFC with serpentine channel flow bed. J. Power Sources 157 (2006), 181.

[162] Lee C.-Y., Wu G.-W., Hsieh C.-L.: In situ diagnosis of micrometallic proton exchange membrane fuel cells using microsensors. J. Power Sources 172 (2007), 363.

[163] Lee C.-Y., Hsieh W.-J., Wu G.-W.: Embedded flexible microsensors in MEA for measuring temperature and humidity in a micro-fuel cell. J. Power Sources 181 (2008), 237.

[164] David N.A., Wild P.M., Hu J., Djilali N.: In-fibre bragg grating sensors for distributed temperature measurement in a polymer electrolyte membrane fuel cell. J. Power Sources 192 (2009), 376.

[165] Basu S., Renfro M.W., Gorgun H., Cetegen B.M.: In situ simultaneous measurements of temperature and water partial pressure in a PEM fuel cell under steady state and dynamic cycling. J. Power Sources 159 (2006), 987.

[166] Bégot S., Kauffmann J.: Estimation of internal fuel cell temperatures from surface temperature measurements. J. Power Sources 178 (2008), 316.

[167] Inman K., Wang X., Sangeorzan B.: Design of an optical thermal sensor for proton exchange membrane fuel cell temperature measurement using phosphor thermometry. J. Power Sources 195 (2010), 4753.

[168] Hakenjos A., Hebling C.: Spatially resolved measurement of PEM fuel cells. J. Power Sources 145 (2005), 307.

[169] Hakenjos A., Muenter H., Wittstadt U., Hebling C.: A PEM fuel cell for combined measurement of current and temperature distribution, and flow field flooding. J. Power Sources 131 (2004), 213.

[170] Lebæk J., Ali S.T., Møller P., Mathiasen C., Nielsen L.P., Kær S.K.: Quantification of in situ temperature measurements on a PBI-based high temperature PEMFC unit cell. Int. J. Hydrogen Energy 35 (2010), 9943.

[171] Siegel C., Bandlamudi G., Heinzel A.: Solid-phase temperature measurements in a HTPEM fuel cell. Int. J. Hydrogen Energy 36 (2011), 12977.

[172] Schulze M., Gülzow E., Schönbauer St., Knöri T., Reissner R.: Segmented cells as tool for development of fuel cells and error prevention/prediagnostic in fuel cell stacks. J. Power Sources 173 (2007), 19.

[173] Sauer D.U., Sanders T., Fricke B., Baumhöfer T., Wippermann K., Kulikovsky A.A., Schmitz H., Mergel J.: Measurement of the current distribution in a direct methanol fuel cell – Confirmation of parallel galvanic and electrolytic operation within one cell. J. Power Sources 176 (2008), 477.

[174] Hartnig Ch., Manke I., Kardjilov N., Hilger A., Grünerbel M., Kaczerowski J., Banhart J., Lehnert W.: Combined neutron radiography and locally resolved current density measurements of operating PEM fuel cells. J. Power Sources 176 (2008), 452.

[175] Pérez L.C., Brandão L., Sousa J.M., Mendes A.: Segmented polymer electrolyte membrane fuel cells – A review. Renewable Sustainable Energy Rev. 15 (2011) 169.

[176] http://www.splusplus.de (last accessed June 2012).

[177] Moser H., Wallnöfer E., Hacker V.: High temperature proton exchange membrane fuel cells – The impact of fuel contaminants and temperature on fuel cell performance. A3PS Conference – Alternative propulsion systems and energy carriers, Vienna, Austria, 2009.

[178] Bergmann A., Kurz T., Gerteisen D., Hebling C.: Spatially resolved impedance spectroscopy in PEM fuel cells up to 200°C. In: Proceeding of the 18th World hydrogen energy conference 2010 – WHEC 2010 parallel sessions book 1 – Fuel cell basics / fuel infrastructures, Jülich: Forschungszentrum Jülich GmbH, Zentralbibliothek, Essen, Germany, 2010.

[179] Lobato J., Cãnizares P., Rodrigo M.A., Pinar F.J., Úbeda D.: Study of flow channel geometry using current distribution measurement in a high temperature PEM fuel cell. J. Power Sources 196 (2011), 4209.

[180] Lobato J., Cãnizares P., Rodrigo M.A., Pinar J.P., Mena E., Úbeda D.: Three-dimensional model of a 50 cm^2 high temperature PEM fuel cell. Study of the flow channel geometry influence. Int. J. Hydrogen Energy 35 (2010), 5510.

[181] Siegel C., Bandlamudi G., Beckhaus P., Burfeind J., Filusch F., Heinzel A.: Segmented current and temperature measurement in a HTPEM fuel cell. In: Proceedings of the 6th symposium on fuel cell modelling and experimental validation, Bad Herrenalb/Karlsruhe, Germany, 2009.

[182] Siegel C., Bandlamudi G., Heinzel A.: Locally resolved measurements in a segmented HTPEM fuel cell with straight flow-fields. Fuel Cells 11 (2011), 489.

[183] Yuan X., Wang H., Sun J.C., Zhang J.: AC impedance technique in PEM fuel cell diagnosis – A review. Int. J. Hydrogen Energy 32 (2007), 4365.

[184] Wu J., Yuan X.Z., Wang H., Blanco M., Martin J.J., Zhang J.: Diagnostic tools in PEM fuel cell research: Part I Electrochemical techniques. Int. J. Hydrogen Energy 33 (2008), 1735.

[185] Andreasen S.J., Vang J.R., Kær S.K.: High temperature PEM fuel cell performance characterisation with CO and CO_2 using electrochemical impedance spectroscopy. Int. J. Hydrogen Energy 36 (2012), 9815.

[186] Andreasen S.J., Jespersen J.L., Schaltz E., Kær S.K.: Characterisation and modelling of a high temperature PEM fuel cell stack using electrochemical impedance spectroscopy. Fuel Cells 9 (2009), 463.

[187] Lobato J., Canizares P., Rodrigo M.A., Linares J.J.: PBI-based polymer electrolyte membranes fuel cells: Temperature effects on cell performance and catalyst stability. Electrochim. Acta 52 (2007), 3910.

[188] Mamlouk M., Scott K.: Analysis of high temperature polymer electrolyte membrane fuel cell electrodes using electrochemical impedance spectroscopy. Electrochim. Acta 56 (2011) 5493.

[189] Hu J., Zhang H., Zhai Y., Liu G., Hu J., Yi B.: Performance degradation studies on H_3PO_4 high temperature PEMFC and one-dimensional numerical analysis. Electrochim. Acta 52 (2006), 394.

[190] Hu J., Zhang H., Gang L.: Diffusion-convection/electrochemical model studies on polybenzimidazole (PBI) fuel cell based on AC impedance technique. Energy Convers. Manage. 49 (2008), 1019.

[191] Schaltz E., Jespersen J.L., Rasmussen P.O.: Development of a 400 w high temperature PEM fuel cell power pack: Equivalent circuit modelling. In: Proceedings of the Fuel Cell Seminar, Honolulu, U.S.A., 2006.

[192] Jespersen J.L., Schaltz E., Kær S.K.: Electrochemical characterization of a PBI-based HT-PEM unit cell. J. Power Sources 191 (2009) 289.

[193] Andreaus B., McEvoy A.J., Scherer G.G.: Analysis of performance losses in polymer electrolyte fuel cells at high current densities by impedance spectroscopy. Electrochim. Acta 47 (2002), 2223.

[194] Brett D.J.L., Atkins S., Brandon N.P., Vesovic V., Vasileiadis N., Kucernak A.R.: Localized impedance mesurements along a single channel of a solid polymer fuel cell. Electrochem. Solid-State Lett. 6 (2003), A63.

[195] Hakenjos A., Zobel M., Clausnitzer J., Hebling C.: Simultaneous electrochemical impedance spectroscopy of single cells in a PEM fuel cell stack. J. Power Sources 154 (2006), 360.

[196] Schneider I.A., Kuhn H., Wokaun A., Scherer G.G.: Study of water balance in a polymer electrolyte fuel cell by locally resolved impedance spectroscopy. J. Electrochem. Soc. 152 (2005), A2383.

[197] Schneider I.A., Kuhn H., Wokaun A., Scherer G.G.: Fast locally resolved electrochemical impedance spectroscopy in polymer electrolyte fuel cells. J. Electrochem. Soc. 152 (2005), A2092.

[198] Schneider I.A., von Dahlen S., Wokaun A., Scherer G.G.: A segmented microstructured flow field approach for submillimeter resolved local current measurement in channel and land areas of a PEFC. J. Electrochem. Soc. 157 (2010), B338.

[199] Schneider I.A., Bayer M.H., von Dahlen S.: Locally resolved electrochemical impedance spectroscopy in channel and land areas of a differential polymer electrolyte fuel cell. J. Electrochem. Soc. 158 (2011), B343.

[200] Hogarth W.H.J., Steiner J., Benziger J.B., Hakenjos A.: Spatially-resolved current and impedance analysis of a stirred tank reactor and serpentine fuel cell flow-field at low relative humidity. J. Power Sources 164 (2007), 464.

[201] Gerteisen D., Mérida W., Kurz T., Lupotto P., Schwager M., Hebling C.: Spatially resolved voltage, current and electrochemical impedance spectroscopy measurements. Fuel Cells 11 (2011), 339.

[202] Gerteisen D., Zamel N., Sadeler C., Geiger F., Ludwig V., Hebling C.: Effect of operating conditions on current density distribution and high frequency resistance in a segmented PEM fuel cell. Int. J. Hydrogen Energy 37 (2012), 7736.

[203] Schneider I.A., Freunberger S.A., Kramer D., Wokaun A., Scherer G.G.: Oscillations in gas channels Part I. The forgotten player in impedance spectroscopy in PEFCs. J. Electrochem. Soc. 154 (2007), B383.

[204] Schneider I.A., Kramer D., Wokaun A., Scherer G.G.: Oscillations in gas channels II. Unraveling the characteristics of the low frequency loop in air-fed PEFC impedance spectra. J. Electrochem. Soc. 154 (2007), B770.

[205] Kulikovsky A.A.: A model for local impedance of the cathode side of PEM fuel cell with segmented electrodes. J. Electrochem. Soc. 159 (2012), F294.

[206] Reshetenko T.V., Bender G., Bethune K., Rocheleau R.: Systematic study of back pressure and anode stoichiometry effects on spatial PEMFC performance distribution. Electrochim. Acta 56, (2011) 8700.

[207] Reshetenko T.V., Bender G., Bethune K., Rocheleau R.: Systematic studies of the gas humidification effects on spatial PEMFC performance distributions. Electrochim. Acta 69 (2012), 220.

[208] Reshetenko T.V., Bender G., Bethune K., Rocheleau R.: Effects of local variations of the gas diffusion layer properties on PEMFC performance using a segmented cell system. Electrochim. Acta 80 (2012), 368.

[209] Reshetenko T.V., Bethune K., Rocheleau R.: Spatial proton exchange membrane fuel cell performance under carbon monoxide poisoning at a low concentration using a segmented cell system. J. Power Sources 218 (2012), 412.

[210] Reshetenko T.V., Bender G., Bethune K., Rocheleau R.: A segmented cell approach for studying the effects of serpentine flow field parameters on PEMFC current distribution. Electrochim. Acta 88 (2013), 571.

[211] Siegel C., Buder I., Bandlamudi G., Heinzel A.: Electrochemical measurements using a segmented HTPEM fuel cell. In: Proceedings of Electrochemistry 2010 (GDCh), Bochum, Germany, 2010.

[212] http://www.frt-gmbh.de (last accessed June 2012).

[213] http://www.agilent.com (last accessed June 2012).

[214] http://www.bader.net/ (last accessed June 2012).

[215] Personal communication with Strunz W., Zahner, Germany, 2012.

[216] http://www.ni.com (last accessed June 2012).